Lecture Notes in Mathematics   1364

Editors:
A. Dold, Heidelberg
B. Eckmann, Zürich
F. Takens, Groningen

Robert R. Phelps

Convex Functions, Monotone Operators and Differentiability

2nd Edition

Springer-Verlag

Berlin Heidelberg New York
London Paris Tokyo
Hong Kong Barcelona
Budapest

Author

Robert R. Phelps
Department of Mathematics GN-50
University of Washington
Seattle, WA 98195-0001, USA

Cover Graphic by Diane McIntyre

Mathematics Subject Classification (1991): 46B20, 46B22, 47H05, 49A29, 49A51, 52A07

ISBN 3-540-56715-1 Springer-Verlag Berlin Heidelberg New York
ISBN 0-387-56715-1 Springer-Verlag New York Berlin Heidelberg

Library of Congress Cataloging-in-Publication Data

Phelps, Robert R. (Robert Ralph), 1926
Convex functions, monotone operators, and differentiability/Robert R. Phelps. -
2nd. ed. p. cm. - (Lecture notes in mathematics; 1364)
Includes bibliographical references and index.
ISBN 0-387-56715-1
1. convex functions. 2. Monotone operators. 3. Differentiable Functions. I. Title.
II. Series: Lecture notes in mathematics (Springer Verlag); 1364.
QA3.L28 no. 1364 (QA331.5) 510 s-dc20 (515'.8)

2146/3140-543210 - Printed on acid-free paper

PREFACE

In the three and a half years since the first edition to these notes was written there has been progress on a number of relevant topics. D. Preiss answered in the affirmative the decades old question of whether a Banach space with an equivalent Gâteaux differentiable norm is a weak Asplund space, while R. Haydon constructed some very ingenious examples which show, among other things, that the converse to Preiss' theorem is false. S. Simons produced a startlingly simple proof of Rockafellar's maximal monotonicity theorem for subdifferentials of convex functions. G. Godefroy, R. Deville and V. Zizler proved an exciting new version of the Borwein-Preiss smooth variational principle. Other new contributions to the area have come from J. Borwein, S. Fitzpatrick, P. Kenderov, I. Namioka, N. Ribarska, A. and M. E. Verona and the author.

Some of the new material and substantial portions of the first edition were used in a one-quarter graduate course at the University of Washington in 1991 (leading to a number of corrections and improvements) and some of the new theorems were presented in the Rainwater Seminar. An obvious improvement is due to the fact that I learned to use TEX. The task of converting the original MacWrite text to TEXwas performed by Ms. Mary Sheetz, to whom I am extremely grateful.

Robert R. Phelps
February 6, 1992
Seattle, Washington

PREFACE TO THE FIRST EDITION

These notes had their genesis in a widely distributed but unpublished set of notes *Differentiability of convex functions on Banach spaces* which I wrote in 1977–78 for a graduate course at University College London (UCL). Those notes were largely incorporated into J. Giles' 1982 Pitman Lecture Notes *Convex analysis with application to differentiation of convex functions.* In the course of doing so, he reorganized the material somewhat and took advantage of any simpler proofs available at that time. I have not hesitated to return the compliment by using a few of those improvements. At my invitation, R. Bourgin has also incorporated material from the UCL notes in his extremely comprehensive 1983 Springer Lecture Notes *Geometric aspects of convex sets with the Radon-Nikodym property.* The present notes do not overlap too greatly with theirs, partly because of a substantially changed emphasis and partly because I am able to use results or proofs that have come to light since 1983.

Except for some subsequent revisions and modest additions, this material was covered in a graduate course at the University of Washington in Winter Quarter of 1988. The students in my class all had a good background in functional analysis, but there is not a great deal needed to read these notes, since they are largely self-contained; in particular, no background in convex functions is required. The main tool is the separation theorem (a.k.a. the Hahn-Banach theorem); like the standard advice given in mountaineering classes (concerning the all-important bowline for tying oneself into the end of the climbing rope), you should be able to employ it using only one hand while standing blindfolded in a cold shower.

These notes have been influenced very considerably by frequent conversations with Isaac Namioka (who has an almost notorious instinct for simplifying proofs) as well as occasional conversations with Terry Rockafellar; I am grateful to them both. I am also grateful to Jon Borwein, Marian Fabian and Simon Fitzpatrick, each of whom sent me useful suggestions based on a preliminary version.

Robert R. Phelps
October 5, 1988
Seattle, Washington

INTRODUCTION

The study of the differentiability properties of convex functions on infinite dimensional spaces has continued on and off for over fifty years. There are a couple of obvious reasons for this. Aside from the intrinsic interest of investigating the many consequences implicit in something as simple as convexity, there is the satisfaction (for this author, at least) in discovering that a number of apparently disparate mathematical topics (extreme points – rather, strongly exposed points – of noncompact convex sets, monotone operators, perturbed optimization of real-valued functions, differentiability of vector-valued measures) are in fact closely intertwined, with differentiability of convex functions forming a common thread.

Starting in Section 1 with the definition of convex functions and a fundamental differentiability property in the one-dimensional case [right-hand and left-hand derivatives always exist], we get quickly to the first infinite dimensional result, Mazur's intriguing 1933 theorem: A continuous convex function on a separable Banach space has a dense G_δ set of points where it is Gâteaux differentiable. In order to go beyond Mazur's theorem, some time is spent in studying the *subdifferential* of a convex function f; this is a set-valued map from the space to its dual whose image at each point x consists of all plausible candidates for the derivative of f at x. [The function f is Gâteaux differentiable precisely when the subdifferential is single-valued, and it is Fréchet differentiable precisely when its subdifferential is single-valued and norm-to-norm continuous.]

Since a subdifferential is a special case of a *monotone operator*, Section 2 starts with a detailed look at monotone operators. These objects are of independent origin, having been extensively studied in the sixties and early seventies by numerous mathematicians (with major contributions from H. Brezis, F. Browder and G. J. Minty) in connection with nonlinear partial differential equations and other aspects of nonlinear analysis. (See, for instance, [Bre], [De], [Pa-Sb] or [Ze]). Also in the sixties, an in-depth study of monotone operators in fairly general spaces was carried out by R. T. Rockafellar, who established a number of fundamental properties, such as their local boundedness. He also gave an elegant characterization of those monotone operators which are the subdifferentials of convex functions. [The connection between monotone operators and derivatives of convex functions is readily apparent on the real line, since single-valued monotone operators coincide in that case with monotone nondecreasing functions, as do the right-hand derivatives of convex functions of one variable.]

Mazur's theorem was revisited 30 years later by J. Lindenstrauss, who showed in 1963 that if a separable Banach space is assumed to be reflexive, then Mazur's conclusion about Gâteaux differentiability could be strengthened to Fréchet differentiability. In 1968, E. Asplund extended Mazur's theorem in two ways: He found more general spaces in which the same conclusion holds, and he studied a smaller class of spaces (now called Asplund spaces) in which Lindenstrauss' Fréchet differentiability conclusion is valid. Asplund used an

ingenious combination of analytic and geometric techniques to prove some of
the basic theorems in the subject. Roughly ten years later, P. Kenderov (as well
as R. Robert and S. Fitzpatrick) proved some general continuity theorems for
monotone operators which, when applied to subdifferentials, yield Asplund's
results as special cases. In Section 2 we follow this approach, incorporating
recent work by D. Preiss and L. Zajicek to obtain the major differentiability
theorems.

The results of Section 2 all involve *continuous* convex functions defined
on open convex sets. For many applications, it is more suitable to consider
lower semicontinuous convex functions, even those which are extended real
valued (possibly equal to $+\infty$). (For instance, in many optimization problems
one finds just such a function in the form of the supremum of an infinite family of affine continuous functions.) Lower semicontinuous convex functions also
yield a natural way to translate results about closed convex sets into results
about convex functions and vice versa. (For instance, the set of points on or
above the graph of such a convex function – its *epigraph* – forms a closed convex set). In Section 3 one will find some classical results (various versions and
extensions of the Bishop-Phelps theorems) which, among other things, guaran-
tee that subdifferentials still exist for lower semicontinuous convex functions.
A nonconvex version of this type of theorem is I. Ekeland's *variational principle*, which asserts that a lower semicontinuous function which nearly attains
its minimum at a point x admits arbitrarily small perturbations (by translates of the norm) which *do* attain a minimum, at points near x. This result,
while simple to state and prove, has been shown by Ekeland [Ek] to have
an extraordinarily wide variety of applications, in areas such as optimization,
mathematical programming, control theory, nonlinear semigroups and global
analysis.

In Section 4, we examine variational principles which use *differentiable*
perturbations. The first such result was due to J. Borwein and D. Preiss;
subsequently, this was recast in a different and somewhat simpler form by
G. Godefroy, R. Deville and V. Zizler; we follow their approach. Some
deep theorems about differentiability of convex functions fall out as fairly
easy corollaries, and it is reasonable to expect future useful applications. This
is followed by the generalization (to maximal monotone operators) of Preiss'
theorem that Gâteaux differentiability of the norm forces *every* continuous
convex function to be generically Gâteaux differentiable.

Section 5 describes the duality between Asplund spaces and spaces with
the *Radon-Nikodym property* (RNP). These are Banach spaces for which a
Radon-Nikodym-type differentiation theorem is valid for vector measures with
values in the space. Spaces with the RNP have an interesting history, starting
in the late sixties with the introduction by M. Rieffel of a geometric property
(*dentability*) which turned out to characterize the RNP and which has led
to a number of other characterizations in terms of the extreme points (or
strongly exposed points) of bounded closed convex subsets of the space. A truly
beautiful result in this area is the fact that a Banach space is an Asplund space
if and only if its dual has the RNP. (Superb expositions of the RNP may be

found in the books by J. Diestel and J. J. Uhl [Di-U] and R. Bourgin [Bou].)
In Section 5, the RNP is defined in terms of dentability, and a number of
basic results are obtained using more recent (and simpler) proofs than are
used in the above monographs. One will also find there J. Bourgain's proof
of C. Stegall's perturbed optimization theorem for semicontinuous functions
on spaces with the RNP; this yields as a corollary the theorem that in such
spaces every bounded closed convex set is the closed convex hull of its strongly
exposed points.

The notion of perturbed optimization has been moving closer to center
stage, since it not only provides a more general format for stating previously
known theorems, but also permits the formulation of more general results. The
idea is simple: One starts with a real-valued function f which is, say, lower
semicontinuous and bounded below on a nice set, and then shows that there
exist arbitrarily small perturbations g such that $f + g$ attains a minimum
on the set. The perturbations g might be restrictions of continuous linear
functionals of small norm, or perhaps Lipschitz functions of small Lipschitz
norm. Moreover, for really nice sets, the perturbed function attains a *strong*
minimum: Every mimimizing sequence converges.

The brief Section 6 is devoted to the class of Banach spaces in which every
continuous convex function is Gâteaux differentiable in a dense set of points
(dropping the previous condition that the set need be a G_δ). Some evidence
is presented that this is perhaps the "right" class to study.

Even more general than monotone operators is a class of set valued maps
(from a metric space, say, to a dual Banach space) which are upper semicontinuous and take on weak* compact convex values, the so-called *usco* maps. In
Section 7, some interesting connections between monotone operators and usco
maps are described, culminating in a topological proof of one of P. Kenderov's
continuity theorems.

CONTENTS

Section 1

1. Convex functions on real Banach spaces.

The letter E will always denote a real Banach space, D will be a nonempty open convex subset of E and f will be a convex function on D. That is, $f: D \to R$ satisfies

$$f[tx + (1-t)y] \leq tf(x) + (1-t)f(y)$$

whenever $x, y \in D$ and $0 < t < 1$. If equality always holds, f is said to be *affine*. A function $f: D \to R$ is said to be *concave* if $-f$ is convex. We will be studying the differentiability properties of such functions, assuming, in the beginning, that they are continuous. (The important case of lower semicontinuous convex functions is considered in Sec. 3.)

1.1. Examples.

(a) The norm function $f(x) = \|x\|$ is an obvious example. More generally, if C is a nonempty convex subset of E, then the distance function

$$d_C(x) = \inf\{\|x - y\|: y \in C\}, \qquad x \in E,$$

is continuous and convex on $D = E$. (Note that $d_C(x) = \|x\|$ if $C = \{0\}$.)

(b) The supremum of any family of convex functions is convex on the set where it is finite. In particular, if A is a nonempty bounded subset of E, then the farthest distance function $x \to \sup\{\|x - y\|: y \in A\}$ is continuous and convex on $D = E$.

(c) The norm function is also generalized by *sublinear functionals*, that is, functions $p: E \to R$ which satisfy

$$p(x + y) \leq p(x) + p(y) \text{ and } p(tx) = tp(x) \text{ whenever } t \geq 0.$$

Obviously, the supremum of a finite family of linear functionals is sublinear. A sublinear functional p is continuous if and only if there exists $M > 0$ such that $p(x) \leq M\|x\|$ for all x.

(d) The *Minkowski gauge functional* is another generalization of the norm function: Suppose that C is a convex subset of E, with $0 \in \text{int } C$. Define

$$p_C(x) = \inf\{\lambda > 0: x \in \lambda C\}, \qquad x \in E.$$

The functional p_C is sublinear and nonnegative. Moreover, $p_C(x) = 0$ if and only if $\{\lambda x: \lambda \geq 0\} \subset C$, and bdry $C = \{x: p_C(x) = 1\}$; in fact

$$\text{int } C = \{x : p_C(x) < 1\} \subset C \subset \{x : p_C(x) \leq 1\} = \overline{C}.$$

There exists $M > 0$ such that $p_C(x) \leq M\|x\|$ for all x (take $M = 1/r$, where the ball of radius r centered at 0 is contained in C), hence p_C is necessarily continuous. Conversely, any positive-homogeneous, subadditive, nonnegative and continuous functional p on E is of the form p_C, simply take $C = \{x : p(x) \leq 1\}$. If p fails to be a seminorm, that is, if $p(x) \neq p(-x)$, then C is not symmetric with respect to 0. Conversely, if C is either open or closed and is not symmetric, then p is not a seminorm.

The following elementary lemma is fundamental to the study of differentiability of convex functions.

Lemma 1.2. *If $x_0 \in D$, then for each $x \in E$ the "right hand" directional derivative*

$$d^+ f(x_0)(x) = \lim_{t \to 0^+} \frac{f(x_0 + tx) - f(x_0)}{t}$$

exists and defines a sublinear functional on E.

Proof. Note that since D is open, $f(x_0 + tx)$ is defined for sufficiently small $t > 0$. Figure 1.1 below shows why $d^+ f(x_0)$ exists; the difference quotient is nonincreasing as $t \to 0^+$, and bounded below, by the corresponding difference quotient from the left.

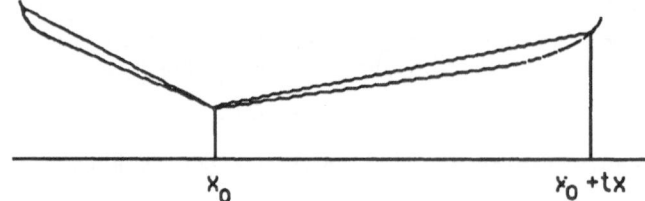

$$x_0 \qquad\qquad\qquad x_0 + tx$$

Fig. 1.1

To prove this, we can assume that $x_0 = 0$ and $f(x_0) = 0$. If $0 < t < s$, then by convexity

$$f(tx) \leq \frac{t}{s} f(sx) + \frac{(s-t)}{s} f(0) = \frac{t}{s} f(sx),$$

which proves monotonicity. Applying this to $-x$ in place of x, we see that

$$-[f(x_0 - tx) - f(x_0)]/t$$

is non*decreasing* as $t \to 0^+$. Moreover, by convexity again, for $t > 0$

$$2f(x_0) \leq f(x_0 - 2tx) + f(x_0 + 2tx),$$

so that

$$\frac{-[f(x_0 - 2tx) - f(x_0)]}{2t} \leq \frac{[f(x_0 + 2tx) - f(x_0)]}{2t}$$

which shows that the right side is bounded below and the left is bounded above. Thus, both limits exist; the left one is $-d^+ f(x_0)(-x)$ and we obviously have

$$-d^+f(x_0)(-x) \leq d^+f(x_0)(x).$$

It is also obvious that $d^+f(x)$ is positively homogeneous. To see that it is subadditive, use convexity again to show that for $t > 0$,

$$\frac{[f(x + t(u + v)) - f(x)]}{t} \leq \frac{f(x + 2tu) - f(x)}{2t} + \frac{f(x + 2tv) - f(x)}{2t}$$

and take limits as $t \to 0^+$.

Definition 1.3. The convex function f is said to be *Gâteaux differentiable* at $x_0 \in D$ provided the limit

$$df(x_0)(x) = \lim_{t \to 0} \frac{f(x_0 + tx) - f(x_0)}{t}$$

exists for each $x \in E$. The function $df(x_0)$ is called the *Gâteaux derivative* (or *Gâteaux differential*) of f at x_0.

It is immediate from this definition (requiring the existence of a two-sided limit) that f is Gâteaux differentiable at x_0 if and only if $-d^+f(x_0)(-x) = d^+f(x_0)(x)$ for each $x \in E$. Since a sublinear functional p is linear if and only if $p(-x) = -p(x)$ for all x, this shows that f is Gâteaux differentiable at x_0 if and only if

$$x \to d^+f(x_0)(x)$$

is linear in x; in particular, if this is true, then $df(x_0)$ is a linear functional on E.

1.4. Examples.

(a) If f is a linear functional on E (not necessarily continuous), then $df(x_0)(x) = f(x)$ for all x_0 and x. For an example of a discontinuous linear functional on a normed linear space, let $f(x) = x'(0)$, for x in the space of all polynomials on $[-1, 1]$ with supremum norm. (It is easy to construct a sequence of polynomials x_n converging uniformly to 0 such that $x_n'(0) = 1$ for all n.) Thus, $x \to df(x_0)(x)$ need not be continuous.

(b) The norm $\|x\| = \Sigma |x_n|$ in ℓ^1 is Gâteaux differentiable precisely at those points $x = (x_n)$ for which $x_n \neq 0$ for all n. In this case, the Gâteaux differential is the bounded sequence $(\text{sgn } x_n) \in \ell^\infty$. The norm in $\ell^1(\Gamma)$ (Γ uncountable) is not Gâteaux differentiable at any point.

Proof. If $x \in \ell^1$ and $x_n = 0$ for some n, let $\delta_n = (0, 0, \ldots, 0, 1, 0, \ldots)$ be the sequence whose only nonzero term is a 1 in the n-th place. It follows that $\|x + t\delta_n\|_1 - \|x\|_1 = |t|$, so dividing both sides by t shows that the (two-sided) limit as $t \to 0$ does not exist. [This observation shows how to prove the second assertion, since any element of $\ell^1(\Gamma)$ vanishes at all but a countable number of members of Γ.] Suppose, on the other hand, that for every n, $x_n \neq 0$, that $\epsilon > 0$ and that $y \in \ell^1$. We can choose $N > 0$ such that $\Sigma_{n>N}|y_n| < \epsilon/2$. For sufficiently small $\delta > 0$ we have

$$\text{sgn}(x_n + ty_n) = \text{sgn } x_n \text{ if } 1 \leq n \leq N, \quad |t| < \delta.$$

Consequently,

$$|\frac{\|x+ty\|_1 - \|x\|_1}{t} - \Sigma y_n \text{ sgn } x_n| \leq$$

$$|\Sigma_{n=1}^{N} t^{-1}\{|x_n+ty_n| - |x_n| - ty_n \text{ sgn } x_n\}| + 2\Sigma_{n>N}|y_n| < \epsilon$$

provided $|t| < \delta$.

If f is a continuous convex function which is Gâteaux differentiable at a point, then its differential is in fact a *continuous* linear functional. This is a consequence of the following basic result.

1.5.   Notation. If $x \in E$ and $r > 0$, the closed ball centered at x is denoted by $B(x;r) = \{y \in E: \|x - y\| \leq r\}$.

Proposition 1.6. *If the convex function f is continuous at $x_0 \in D$, then it is locally Lipschitzian at x_0, that is, there exist $M > 0$ and $\delta > 0$ such that $B(x_0;\delta) \subset D$ and*

$$|f(x) - f(y)| \leq M\|x - y\|$$

whenever $x, y \in B(x_0;\delta)$.

Proof. Since f is continuous at x_0, it is locally bounded there; that is, there exist $M_1 > 0$ and $\delta > 0$ such that $|f| \leq M_1$ on $B(x_0;2\delta) \subset D$. If x, y are distinct points of $B(x_0;\delta)$, let $\alpha = \|x - y\|$ and let

$$z = y + (\delta/\alpha)(y - x);$$

see Fig. 1.2 below.

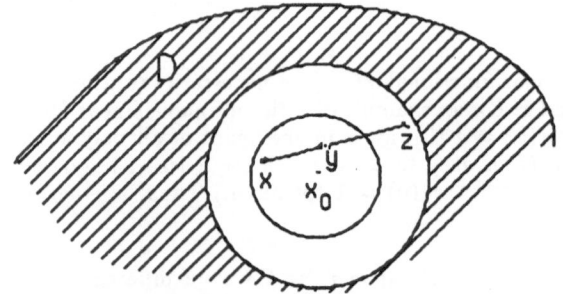

Fig. 1.2

Note that $z \in B(x_0;2\delta)$. Since $y = [\alpha/(\alpha + \delta)]z + [\delta/(\alpha + \delta)]x$ is a convex combination (lying in $B(x_0;2\delta)$), we have

$$f(y) \leq [\alpha/(\alpha + \delta)]f(z) + [\delta/(\alpha + \delta)]f(x) \qquad \text{so}$$

$$f(y) - f(x) \leq [\alpha/(\alpha + \delta)]\{f(z) - f(x)\} \leq (\alpha/\delta) \cdot 2M_1 = (2M_1/\delta)\|x - y\|.$$

Interchanging x and y gives the desired result, with $M = 2M_1/\delta$.

Remark. Note that we only used local boundedness of f; hence the latter property is *equivalent* to continuity for a convex function on an open convex set. In fact, it suffices that f be merely locally bounded above: If $f \leq M$ on the ball $B(x_0; r)$, then for all $x \in B(x_0; r)$ we have $2x_0 - x \in B(x_0; r)$ and hence

$$f(x_0) \leq \frac{f(2x_0 - x) + f(x)}{2} \leq \frac{M + f(x)}{2},$$

so $-f(x) \leq M + 2|f(x_0)|$; that is, $|f| \leq M + 2|f(x_0)|$ on $B(x_0; r)$.

Corollary 1.7. *If the convex function f is continuous at $x_0 \in D$, then $d^+f(x_0)$ is a continuous sublinear functional on E, and hence $df(x_0)$ (when it exists) is a continuous linear functional.*

Proof. Given $x_0 \in D$ there exists a neighborhood B of x_0 and $M > 0$ such that, if $x \in E$, then

$$f(x_0 + tx) - f(x_0) \leq Mt\|x\|$$

provided $t > 0$ is sufficiently small that $x_0 + tx \in B$. Thus, for all points $x \in E$, we have $d^+f(x_0)(x) \leq M\|x\|$, which implies that $d^+f(x_0)$ is continuous.

Proposition 1.8. *The continuous convex function f is Gâteaux differentiable at $x_0 \in D$ if and only if there exists a unique functional x^* in E^* satisfying*

$$\langle x^*, x - x_0 \rangle \leq f(x) - f(x_0), \qquad x \in D, \tag{1.1}$$

or equivalently

$$\langle x^*, y \rangle \leq d^+f(x_0)(y), \qquad y \in E. \tag{1.2}$$

Proof. We first show that (1.1) and (1.2) are equivalent. If x^* satisfies (1.1), then for any $y \in E$ we have $x_0 + ty \in D$ for sufficiently small $t > 0$ hence $t\langle x^*, y \rangle = \langle x^*, (x_0 + ty) - x_0 \rangle \leq f(x_0 + ty) - f(x_0)$ which implies that x^* satisfies (1.2). Conversely, if x^* satisfies (1.2) and $x \in D$, let $y = x - x_0$; then $x_0 + t(x - x_0) \in D$ if $0 < t \leq 1$ so

$$\langle x^*, x - x_0 \rangle \leq d^+f(x_0)(x - x_0) \leq t^{-1}[f(x_0 + t(x - x_0)) - f(x_0)].$$

Setting $t = 1$ yields (1.1).

If $df(x_0)$ exists, then $df(x_0)(x - x_0) \leq f(x) - f(x_0)$ as above, so $df(x_0)$ satisfies (1.1). Moreover, if x^* satisfies (1.1), then it satisfies (1.2); linearity of $d^+f(x_0) = df(x_0)$ implies that $x^* = df(x_0)$.

Conversely, suppose that x^* is the unique element of E^* satisfying (1.1), hence the unique element satisfying (1.2). We now apply the general fact that if a continuous sublinear functional p majorizes exactly one linear functional, then p is itself linear. Indeed, if p is not linear, then it dominates many linear functionals (see the sketch below); the proof is an easy consequence of the Hahn-Banach theorem: If $-p(-x) < p(x)$, find p-dominated extensions of the linear functionals

$$f_1(rx) = rp(x) \qquad \text{and} \qquad f_2(rx) = -rp(-x).$$

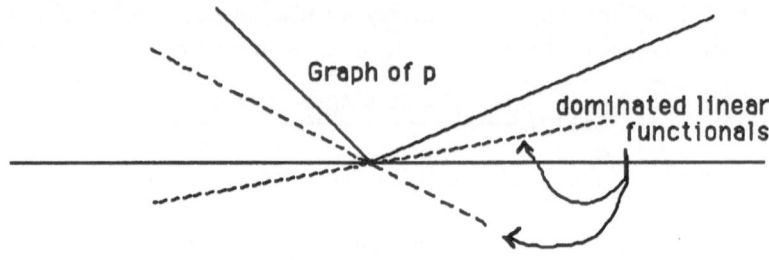

Fig. 1.3

The functionals x^* which satisfy (1.1) play an important role in the study of convex functions, so they are singled out for special attention.

Definition 1.9. If f is a convex function defined on the convex set C and $x \in C$, we define the *subdifferential of f at x* to be the set $\partial f(x)$ of all $x^* \in E^*$ satisfying

$$\langle x^*, y - x \rangle \leq f(y) - f(x) \text{ for all } y \in C.$$

Note that this is the same as saying that the affine function $x^* + \alpha$, where $\alpha = f(x) - \langle x^*, x \rangle$, is dominated by f and is equal to it at $y = x$, as indicated in the sketch.

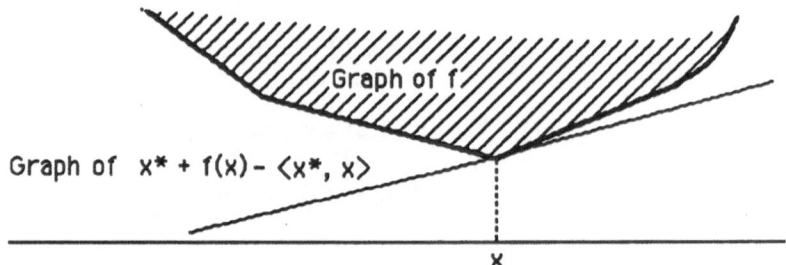

Fig. 1.4

The Hahn-Banach argument we used above shows quickly that if f is continuous at x_0, then $\partial f(x_0)$ is nonempty: $d^+ f(x_0)$ is continuous and sublinear, so (as above) there exists x^* such that $\langle x^*, y \rangle \leq d^+ f(x_0)(y)$ for all $y \in E$. Using the fact that the right-hand difference quotients for $d^+ f(x_0)$ are decreasing, replacing y by $y - x_0$ and letting $t = 1$, we get

$$\langle x^*, y - x_0 \rangle \leq d^+ f(x_0)(y - x_0) \leq f((y - x_0) + x_0) - f(x_0) \text{ for all } y \in C.$$

As we will see later, it is still possible to have $\partial f(x_0)$ nonempty for certain convex f which are *not* continuous at x_0.

1.10. Exercise.

Prove that for any convex function f the set $\partial f(x_0)$ (possibly empty!) is convex and weak* closed. (Note that a continuous convex f is Gâteaux differentiable at x_0 if and only if $\partial f(x_0)$ is a singleton.)

Proposition 1.11. *If the convex function f is continuous at $x_0 \in D$, then $\partial f(x_0)$ is a nonempty, convex and weak* compact subset of E^*. Moreover, the map $x \to \partial f(x)$ is locally bounded at x_0, that is, there exist $M > 0$ and neighborhood U of x_0 in D such that $\|x^*\| \leq M$ whenever $x \in U$ and $x^* \in \partial f(x)$.*

Proof. The fact that $\partial f(x_0)$ is nonempty, weak* closed and convex follows from the preceding remarks and Exercise 1.10. The fact that it is weak* compact will follow from Alaoglu's theorem, once we have shown the local boundedness property. Since, by Proposition 1.6, f is locally Lipschitzian at x_0, there exist $M > 0$ and a neighborhood U of x_0 such that

$$|f(y) - f(x)| \leq M\|y - x\| \text{ whenever } x, y \in U.$$

If $x \in U$ and $x^* \in \partial f(x)$, then for all $y \in U$ we have

$$\langle x^*, y - x \rangle \leq f(y) - f(x) \leq M\|y - x\|,$$

which implies that $\|x^*\| \leq M$.

Definition 1.12. Suppose that E and F are normed linear spaces, that U is a nonempty open subset of E and that $\varphi : U \to F$ is a continuous function. We can extend the definition of *Gâteaux differentiability* as follows: Say that φ is Gâteaux differentiable at the point $x_0 \in U$ provided there exists a continuous linear map from E to F (denoted by $d\varphi(x_0)$) such that

$$d\varphi(x_0)(x) = \lim_{t \to 0+} t^{-1}\{\varphi(x_0 + tx) - \varphi(x_0)\} \text{ for each } x \in E. \qquad (1.3)$$

Another way of stating this is to say that φ has directional derivatives at x_0 in every direction x and the resulting function of x is continuous and linear.

We say that φ is *Fréchet differentiable* at $x_0 \in U$ provided there exists a continuous linear map from E to F (denoted by $\varphi'(x_0)$) such that for all $\epsilon > 0$, there exists $\delta > 0$ such that

$$\|\varphi(x_0 + x) - \varphi(x_0) - \varphi'(x_0)(x)\| \leq \epsilon\|x\| \text{ whenever } \|x\| < \delta. \qquad (1.4)$$

We call $\varphi'(x_0)$ (which is easily seen below to be unique) the *Fréchet differential* (or *Fréchet derivative*) of φ.

For the moment, we will be dealing with *real*-valued continuous functions, so Gâteaux and Fréchet derivatives will be continuous linear maps from E into $\mathbf{R}$, that is, elements of E^*.

1.13 Facts.

(a) If f is a continuous function which is Fréchet differentiable at x_0, then it is Gâteaux differentiable there and $\phi'(x_0) = d\varphi(x_0)$. To see this, replace x in (1.4) by tx, fix x and let $t \to 0^+$. Since limits are unique, the operator $d\varphi(x_0)$ is uniquely determined, hence $\varphi'(x_0)$ is unique.

(b) Note that φ is Fréchet differentiable at x_0 if it is Gâteaux differentiable there and if the limit in (1.3) exists *uniformly* for $\|x\| \leq 1$ as $t \to 0^+$.

1.14 Examples.

(a) The norm in ℓ^1 is not Fréchet differentiable at any point.

Proof. By Example 1.4(b), we need only consider a point $x = (x_n)$ for which $x_n \neq 0$ for all n. Given such an x, for each $m \geq 1$ let

$$y^m = (0, 0, ..., 0, -2x_m, -2x_{m+1}, -2x_{m+2}, ...).$$

Then $\|y^m\|_1 \to 0$ as $m \to \infty$. Of course, the sequence (sgn x_n) is our only candidate for the Fréchet differential. But

$$\left| \|x + y^m\|_1 - \|x\|_1 - \Sigma(y^m)_n \text{ sgn } x_n \right| = |\Sigma_{n \geq m}(-2|x_n|)| = \|y^m\|_1.$$

(b) The square of the norm in Hilbert space H is everywhere Fréchet differentiable. By the chain rule, the norm is therefore differentiable at every point other than the origin.

Proof. If $x, y \in H$, then $\|x + y\|^2 - \|x\|^2 - 2(x, y) = \|y\|^2$; it follows readily that $y \to 2(x, y)$ is the Fréchet derivative of $\|\cdot\|^2$ at x.

(c) There exists an equivalent norm on ℓ^1 which is Gâteaux differentiable at every point (except the origin), but is *nowhere* Fréchet differentiable. This striking example will be easy to prove after we have developed a few tools in later sections, so it will be postponed until Sec. 5 (following Theorem 5.12).

(d) In Hilbert space H let C be a nonempty closed convex set and denote by P the Lipschitz continuous *nearest point mapping* (or *metric projection*) of H onto C; that is, for all $x \in H$, $P(x)$ is the unique point satisfying

$$\|x - P(x)\| = \inf\{\|x - y\| : y \in C\}.$$

Define f on H by

$$f(x) = (\tfrac{1}{2})[\|x\|^2 - \|x - P(x)\|^2];$$

then f is continuous, convex and everywhere Fréchet differentiable, with $f'(x) = P(x)$ for all x.

Proof. We first prove the fundamental fact that the mapping P satisfies (in fact, it is characterized by) the following *variational inequality*: For all $x \in H$,

$$\langle x - P(x), z - P(x) \rangle \leq 0 \quad \text{for all } z \in C. \tag{1.5}$$

Indeed, if $z \in C$ and $0 < t < 1$, then $z_t \equiv tz + (1-t)P(x) \in C$ and hence $\|x - P(x)\| \le \|x - z_t\| = \|(x - P(x)) - t(z - P(x))\|$. Squaring both sides of this inequality, expanding and then cancelling $\|x - P(x)\|^2$ on both sides yields

$$0 \le -2t\langle x - P(x), z - P(x)\rangle + t^2\|z - P(x)\|^2.$$

If we then divide by t and take the limit as $t \to 0$ we obtain (1.5). Moreover, if $y \in H$ and we write down (1.5) again, using y in place of x, then take $z = P(y)$ in the first equation, $z = P(x)$ in the second one and add the two, we obtain

$$\langle x - y, P(x) - P(y)\rangle \ge \|P(x) - P(y)\|^2 \quad \text{for all } x, y \in H, \tag{1.6}$$

Since $\langle x - y, P(x) - P(y)\rangle \le \|x - y\| \cdot \|P(x) - P(y)\|$, we see that P is a *contraction*: $\|P(x) - P(y)\| \le \|x - y\|$ for all $x, y \in H$. Returning now to the function f, note that since

$$2f(x) = \|x\|^2 - \inf\{\|x - y\|^2 : y \in C\} = \sup\{2\langle x, y\rangle - \|y\|^2 : y \in C\},$$

it is the supremum of affine functions, hence is convex (and it is clearly continuous). To see the differentiability property, fix $x \in H$; then for any $y \in H$ we have

$$\|(x + y) - P(x + y)\| \le \|(x + y) - P(x)\|, \text{so}$$

$$\|(x + y) - P(x + y)\|^2 \le \|x + y\|^2 - 2\langle x + y, Px\rangle + \|P(x)\|^2$$
$$= \|x + y\|^2 + \|x - P(x)\|^2 - \|x\|^2 - 2\langle y, P(x)\rangle,$$

hence $f(x + y) - f(x) - \langle P(x), y\rangle \ge 0$. On the other hand, since $\|x - P(x)\| \le \|x - P(x + y)\|$ we get

$$f(x + y) - f(x) - \langle P(x), y\rangle \le \langle y, P(x + y) - P(x)\rangle \le$$
$$\le \|y\| \cdot \|P(x + y) - P(x)\| \le \|y\|^2,$$

which implies the differentiability assertion.

1.15. Exercises.

(a) Prove that for continuous convex functions in finite dimensional spaces, Gâteaux differentiability implies Fréchet differentiability. (Hint: Use the Fact 1.13 (b), the local Lipschitz property and compactness of the unit ball in finite dimensional spaces.)

(b) (A calculus student's delight.) In $\mathbf{R}^n$, Gâteaux (hence Fréchet) differentiability of a continuous convex function f at a point x_0 is equivalent to the existence of the partial derivatives $(\partial f / \partial x_i)(x_0)$, $i = 1, 2, 3, \ldots, n$. (Hint: With basis $\{e_i\}$, use the linearity of $d^+ f(x_0)$ on each line $\mathbf{R}e_i$ to show that $\partial f(x_0)$ contains only the single linear functional $x \to \Sigma x_i \frac{\partial f}{\partial x_i}(x_0)$.)

Convex functions on the real line have many points of differentiability, as shown by the following result. Despite its elementary nature, it lies at the heart of the proof of Mazur's theorem for separable Banach spaces (Theorem 1.20, below).

Theorem 1.16. *If f is convex on an open interval $D \subset \mathbf{R}$, then $f'(x)$ exists for all but (at most) countably many points of D.*

Proof. We first show that $d^+f(x)(1)$ [for simplicity, we will write $d^+f(x)$] is a nondecreasing function of x. Suppose, then, that $x_1 < x_2$; we want $d^+f(x_1) \leq d^+f(x_2)$. Without loss of generality, we may assume that $x_1 = 0$ and $f(x_1) = 0$. [If necessary, we can translate both D and f.] Since $d^+f(0) \leq f(x_2)/x_2$, it suffices to show that the latter expression is dominated by $[f(x_2+t)-f(x_2)]/t$ whenever $t > 0$. But if we take $\lambda = x_2/(x_2 + t)$, then we have $x_2 = \lambda(x_2 + t) + (1 - \lambda)0$, so by convexity,

$$f(x_2) \leq \lambda f(x_2 + t),$$

which is equivalent to the desired inequality. A geometrical proof of the monotonicity of d^+f can be obtained from the following sketch:

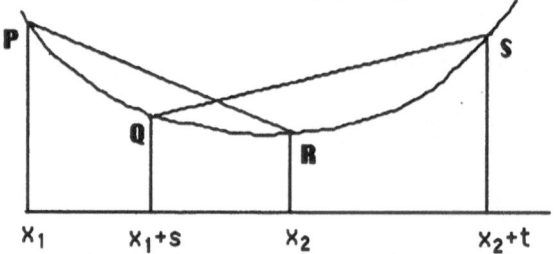

Fig. 1.5

Denoting the various chords to the graph of f by PQ, QR, etc., it is clear that slope $PQ \leq$ slope $PR \leq$ slope $QR \leq$ slope $QS \leq$ slope RS. Expressing the first and last of these in terms of f, we see that (for any $s > 0$ such that $x_1 + s < x_2$, and any $t > 0$)

$$[f(x_1 + s) - f(x_1)]/s \leq [f(x_2 + t) - f(x_2)]/t,$$

which shows that $d^+f(x_1) \leq d^+f(x_2)$. We next show that any point where f fails to be differentiable is a point where the monotone function $x \to d^+f(x)$ has a jump. There are, of course, at most countably many such points. Now, if $f'(x_0)$ fails to exist, then

$$-d^+f(x_0)(-1) < d^+f(x_0)(1),$$

so it suffices to show that the latter inequality implies that $d^+f(x)$ has a jump at $x = x_0$, that is, that

$$\lim_{x \to x_0^-} d^+f(x) < \lim_{x \to x_0^+} d^+f(x).$$

Since the right side of this expression dominates $d^+f(x_0)$, it suffices to show that the left side is dominated by $-d^+f(x_0)(-1)$, that is, if $x < x_0$, then $d^+f(x)(1) < -d^+f(x_0)(-1)$. In view of the monotonicity of the limits which define these two quantities, we need only show that, letting $t_0 = (\frac{1}{2})(x_0 - x)$ we get $[f(x + t_0) - f(x)]/t_0 \leq [f(x_0 - t_0) - f(x_0)]/t_0$. But this is easily seen to be equivalent to the convexity inequality

$$f[\frac{1}{2}(x + x_0)] \leq \frac{1}{2}[f(x) + f(x_0)],$$

and the proof is complete.

The above conclusion no longer holds in $\mathbf{R}^2$; for instance, the function $(x_1, x_2) \to |x_1|$ fails to be differentiable at each point of the x_2-axis. It *is* differentiable almost everywhere (Lebesgue), and this is generally true in $\mathbf{R}^n$, as shown in the next exercise. Notice that we obtained a differentiability property without assuming f to be continuous. This suggests that continuity of convex functions is automatic; indeed, as we show below, this is true in finite dimensional spaces. The existence of discontinuous linear functionals on any infinite dimensional normed linear space shows that convexity does not imply continuity in the general case.

1.17. Exercise.

Prove that a continuous convex function f on an open convex subset D of $\mathbf{R}^n$ is differentiable almost everywhere. (Hint: Show that for fixed basis element e_k, the function $x \to d^+ f(x)(e_k)$ is a pointwise limit of continuous functions, hence is Borel measuraable and there the set B of points $x \in D$ where $\frac{\partial f}{\partial x_k}(x)$ does not exist is a Borel set. Use Fubini's theorem and Theorem 1.16 to show that B has measure zero. Then use Exercise 1.15(b).)

A more general (and more difficult) result than this is given by a theorem of Rademacher; for perhaps the simplest proof, see [S-P].

Theorem 1.18 (Rademacher). *Let U be a nonempty open subset of $\mathbf{R}^n$ and suppose that $f: U \to \mathbf{R}^m$ is locally Lipschitzian. Then f is (Fréchet) differentiable almost everywhere (Lebesgue).*

Proposition 1.19. *Suppose that f is convex on a nonempty open convex subset D of $\mathbf{R}^n$. Then f is continuous at each point of D.*

Proof. It suffices to assume that $0 \in D$ and to show that f is continuous at 0. By the remark following Prop. 1.5 it suffices to show that f is bounded above in some neighborhood of 0. We can assume that the topology of $\mathbf{R}^n$ is given by the ℓ^1 norm: $\|x\| = \Sigma_{1 \leq k \leq n} |x_k|$. Since D is open, there exists $r > 0$ such that $x \in D$ whenever $\|x\| \leq r$. Letting e_k denote the k-th coordinate vector, $k = 1, 2, \ldots, n$, we can express any such x as $x = \Sigma x_k e_k = \Sigma (x_k/r)(re_k)$, where $\Sigma |x_k/r| = \|x/r\| \leq 1$. Thus, we can write

$$x = \lambda_0 \cdot 0 + \Sigma \lambda_k (re_k) + \Sigma \mu_k (-re_k),$$

where $\lambda_k \geq 0$, $k = 0, 1, 2, \ldots, n$ and $\mu_k \geq 0$, $k = 1, 2, \ldots, n$ and $\Sigma \lambda_k + \Sigma \mu_k = 1$. By the convexity of f, $f(x) \leq \lambda_0 f(0) + \Sigma \lambda_k f(re_k) + \Sigma \mu_k f(-re_k)$, which implies that $f(x) \leq M \equiv \max\{|f(0)|, |f(re_k)|, |f(-re_k)|\}$ in the ball $B(0; r)$.

We now return to our primary interest, the fact that continuous convex functions on certain Banach spaces are necessarily generically (= "in a dense

G_δ subset") differentiable. The following theorem of Mazur motivated much subsequent work in this direction. As we will see, it reduces the problem to the one-dimensional theorem.

Theorem 1.20 (Mazur, 1933). *If E is a separable Banach space and f is a continuous convex function defined on a convex open subset D of E, then the set of points x where $df(x)$ exists is a dense G_δ set in D.*

Proof. We first show that the set of $x \in D$ where $df(x)$ does not exist is a (relative) F_σ subset of D. Let $\{x_n\}$ be a sequence which is dense in the unit sphere of E and for each $n \geq 1$, $m \geq 1$ let $A_{n,m}$ denote all those x in D for which there exist $x^*, y^* \in \partial f(x)$ satisfying

$$\langle x^* - y^*, x_n \rangle \geq 1/m.$$

Since $df(x)$ doesn't exist if and only if $\partial f(x)$ contains more than one point, it is clear that $df(x)$ fails to exist if and only if $x \in \cup A_{n,m}$. To see that each $A_{n,m}$ is relatively closed, suppose that $\{z_k\} \subset A_{n,m}$ with $z_k \to z$, where $z \in D$. For each k we can choose x_k^* and y_k^* in $\partial f(z_k)$ such that $\langle x_k^* - y_k^*, z_k \rangle \geq 1/m$. Since E is separable, the bounded subsets of E^* are metrizable in the weak* topology, so by local boundedness and weak* compactness there is no loss of generality in assuming that there exist x^* and y^* such that $x_k^* \to x^*$ and $y_k^* \to y^*$ (weak*). It follows that for any $y \in D$, we must have

$$\langle x^*, y - z \rangle = \lim \langle x_k^*, y - z_k \rangle \leq \lim [f(y) - f(z_k)] = f(y) - f(z),$$

so that x^* (and similarly, y^*) is in $\partial f(z)$. Since

$$\langle x^* - y^*, x_n \rangle = \lim \langle x_k^* - y_k^*, x_n \rangle \geq 1/m,$$

we see that $z \in A_{n,m}$. Finally, to see that $D \setminus A_{n,m}$ is dense in D for each n, m, suppose that $x_0 \in D$. By Theorem 1.16, the function

$$f_1(r) = f(x_0 + rx_n) \text{ defined on } I = \{r \in \mathbf{R} : x_0 + rx_n \in D\}$$

is differentiable with the exception of at most countably many points. In particular, we can approximate x_0 by points of the form $x' = x_0 + rx_n$ where $f_1'(r)$ exists. If $x^*, y^* \in \partial f(x')$, then their restrictions to the line $x_0 + \mathbf{R}x_n$ yield subdifferentials to f_1 at x'. Differentiability of f_1 at x' implies that these two restrictions must coincide on the entire line, hence $\langle x^*, x_n \rangle = \langle y^*, x_n \rangle$. It follows that $x' \in D \setminus A_{n,m}$, for $m = 1, 2, 3, \ldots$. We conclude that $\cap (D \setminus A_{n,m})$ is a dense G_δ subset of D, since open subsets of a Banach space have the Baire property.

It has been shown [Č-K] that in certain nonseparable $C(X)$ spaces (for instance, if X is the Čech compactification of the rationals) the set G of points where the supremum norm is Gâteaux differentiable need not be a G_δ; indeed, G can be dense yet not even contain a dense G_δ. While these examples are complicated, the following simple example shows that *some* hypotheses are required for the validity of a generic differentiability theorem.

1.21. Example.

For $x = (x_n)$ in ℓ^∞ define a seminorm p by

$$p(x) = \limsup |x_n|;$$

then p is continuous, but nowhere Gâteaux differentiable.

Proof. Clearly, $p(x) \leq \|x\|_\infty$, so p is a continuous seminorm. If $p(x) = 0$, then $x_n \to 0$, so taking $y = (1,1,1,\ldots)$ we have

$$t^{-1}[p(x+ty) - p(x)] = |t|/t,$$

which shows that $dp(x)$ does not exist. If $p(x) > 0$, we can assume that $p(x) = 1$. Choose a subsequence $\{x_{n(i)}\}$ of $\{x_n\}$ such that $|x_{n(i)}| \to 1$. By taking a further subsequence we can assume that the $x_{n(i)}$ have the same sign and, since $p(x) = p(-x)$, it suffices to consider the case $x_{n(i)} > 0$ for all i. Define $y_n = 0$ if either $n \neq n(i)$ or $n = n(i)$ with i odd, while $y_n = 1$ if $n = n(i)$ with i even. Then

$$t^{-1}[p(x+ty) - p(x)] = 1 \text{ if } t > 0, = 0 \text{ if } -1 < t < 0.$$

Despite the foregoing example, there are nonseparable spaces in which the conclusion to Mazur's theorem remains valid, for instance, the class of *weakly compactly generated* (WCG) Banach spaces (to which we will return later). The attempt to characterize those spaces in which convex continuous functions are always generically differentiable has motivated the following terminology.

Definition 1.22. A Banach space E is said to be an *Asplund space* (*weak Asplund space*) provided every continuous convex function defined on a nonempty open convex subset D of E is Fréchet differentiable (Gâteaux differentiable) at each point of some dense G_δ subset of D.

Note that the term "weak" does not refer to the weak topology; it arose because Gâteaux differentiability has sometimes been called "weak differentiability" (since it is weaker than Fréchet – sometimes called "strong"– differentiability).

Some 30 years after Mazur's paper, J. Lindenstrauss [Li] obtained the next result of this nature, when he showed that if a separable Banach space was also *reflexive*, then the Gâteaux differentiability in Mazur's theorem could be changed from Gâteaux to Fréchet. Five years later, E. Asplund renewed the study of such questions [Asp], making some impressive contributions. Mazur's theorem can be restated in the form "Separable Banach spaces are weak Asplund spaces" (while Example 1.21 shows that ℓ^∞ is not a weak Asplund space and Example 1.14(a) shows that ℓ^1 is not an Asplund space). We will eventually prove, among other things, the fundamental result (due to Asplund) which states that *if E^* is separable, then E is an Asplund space* ; this clearly implies Lindenstrauss' theorem.

For convex continuous functions f, it is possible (and quite useful) to characterize Fréchet differentiability at x solely in terms of f, that is, without mentioning the linear functional $f'(x)$. (This is possible since there are always candidates for the derivative lurking in $\partial f(x)$.)

Proposition 1.23. *Suppose that f is continuous and convex on the nonempty open convex subset D of E. Then f is Fréchet differentiable at $x \in D$ if and only if for all $\epsilon > 0$ there exists $\delta > 0$ such that*

$$f(x + ty) + f(x - ty) - 2f(x) < t\epsilon \qquad (1.7)$$

whenever $\|y\| = 1$ and $0 < t < \delta$.

Proof. If $f'(x)$ exists, then for all $\epsilon > 0$ there exists $\delta > 0$ such that

$$f(x + ty) - f(x) - \langle f'(x), ty \rangle < (\epsilon/2)t\|y\|$$

whenever $0 < t < \delta$ and $\|y\| = 1$. Writing this down with $-y$ in place of y and adding both expressions yields (1.7). Conversely, suppose the stated condition holds. Choose $x^* \in \partial f(x)$; it follows that for all y and all sufficiently small $t > 0$ such that $x \pm ty \in D$,

$$\langle x^*, ty \rangle = \langle x^*, (x + ty) - x \rangle \leq f(x + ty) - f(x) \text{ and} \qquad (1.8)$$

$$-\langle x^*, ty \rangle \leq f(x - ty) - f(x). \qquad (1.9)$$

By hypothesis, for any $\epsilon > 0$ there exists $\delta > 0$ such that (1.7) holds for any $0 < t < \delta$ and any y with $\|y\| = 1$, that is,

$$f(x + ty) - f(x) - \langle x^*, ty \rangle \leq t\epsilon + f(x) - f(x - ty) - \langle x^*, ty \rangle.$$

Inequality (1.8) shows that the left side is greater or equal to 0 while (1.9) shows that the right side is less than or equal to $t\epsilon$, which completes the proof.

1.24. Exercise.

Prove that a continuous convex function f on the open convex set D is Gâteaux differentiable at $x \in D$ if and only if for each $y \in E$

$$\lim\nolimits_{t \to 0+} \frac{f(x + ty) + f(x - ty) - 2f(x)}{t} = 0.$$

The previous proposition makes it easy to formulate and prove the following basic fact.

Proposition 1.25. *Suppose that f is continuous and convex on the nonempty open convex subset D of E. Then the set G (possibly empty) of points $x \in D$ where f is Fréchet differentiable is a G_δ.*

Proof. For each n let G_n be the set of all those $x \in D$ for which there exists $\delta > 0$ such that

$$\sup \frac{f(x + \delta y) + f(x - \delta y) - 2f(x)}{\delta} < 1/n$$

the supremum being taken over all y with $\|y\| = 1$. Recall from Lemma 1.2 that for fixed x and y the functions $t \to t^{-1}[f(x + t(\pm y)) - f(x)]$ are non-increasing as $t \to 0^+$, hence from Prop. 1.23 we can conclude that $G = \cap G_n$. Thus, it suffices to prove that each G_n is open. Suppose, then, that $x \in G_n$. Since f is locally Lipschitzian, there exist $\delta_1 > 0$ and $M > 0$ such that $|f(u) - f(v)| \le M\|u - v\|$ whenever $u, v \in B(x; \delta_1)$. Since $x \in G_n$, there exist $\delta > 0$ and $r > 0$ such that for all y with $\|y\| = 1$ we have $x \pm \delta y \in D$ and

$$\frac{f(x + \delta y) + f(x - \delta y) - 2f(x)}{\delta} \le r < 1/n.$$

Suppose, now, that $z \in B(x; \delta_2) \subset D$, where $0 < \delta_2 < \delta_1$ is sufficiently small that $z \pm \delta_2 y \in D$ for all y with $\|y\| = 1$; a further restriction on δ_2 will be indicated below. Then for any $\|y\| = 1$ we have

$$\delta^{-1}[f(z + \delta y) + f(z - \delta y) - 2f(z)] \le$$

$$\delta^{-1}[f(x + \delta y) + f(x - \delta y) - 2f(x)] + \delta^{-1}|f(z + \delta y) - f(x + \delta y)| +$$

$$+ \delta^{-1}|f(z - \delta y) - f(x - \delta y)| + 2\delta^{-1}|(f(z) - f(x)|$$

$$\le r + 4M\delta^{-1}\|z - x\| \le r + 4M\delta^{-1}\delta_2.$$

Thus, it suffices that δ_2 be chosen small enough that $r + 4M\delta^{-1}\delta_2 < 1/n$.

The usefulness of the foregoing proposition is evident: To prove generic Fréchet differentiability, we need only show that the set G is dense; it is automatically a G_δ. As we noted earlier, in order to prove that a continuous convex function f is generically Gâteaux differentiable, it suffices to show that there is a dense G_δ subset of points x for which $\partial f(x)$ is a singleton. This indicates that it is worthwhile to know more about the set-valued map $x \to \partial f(x)$, which has a number of special properties. Some of these have already been seen: Recall that in Prop. 1.11 we showed that the sets $\partial f(x)$ are nonempty, weak* compact and convex, and that the map ∂f is itself locally bounded. We'll see further fundamental properties in the next section. We conclude this section with an easy observation relating ∂f to optimization.

Proposition 1.26. *A continuous convex function f on a nonempty open subset D has a global minimum at $x \in D$ if and only if $0 \in \partial f(x)$.*

Proof. This is immediate from the definition of $\partial f(x)$.

Remarks.

The material in this introductory section is classical; it provides the background about convex functions and their derivatives needed for the remainder of the notes. Flett's book [Fl] is frequently useful when one needs to verify fundamental questions concerning differentiability and Roberts-Varberg [R-V] is a good source for basic elementary facts about convex functions. Rademacher's theorem (stated as Theorem

1.18) has been extended in various ways to infinite dimensional spaces. The first requirement is to come up with a reasonable definition of "almost everywhere" (that is, of some notion of "measure zero") for infinite dimensional spaces, where no countably additive measure with decent properties can exist. Once this is done, theorems can be proved which assert that a locally Lipschitzian map from a separable Banach space into a space with the RNP (see Sec. 5) must be Gâteaux differentiable almost everywhere. It has been known for at least thirty years that no such theorem holds for Fréchet differentiability; there exist simple locally Lipschitzian maps of separable Hilbert space into itself which are nowhere Fréchet differentiable. There are even several examples (with subsequent citations) which purport to exhibit *real-valued* locally Lipschitzian functions on Hilbert space which are nowhere Fréchet differentiable. In preparing our UCL notes, we found all such examples to be flawed, leaving the obvious question as to whether such functions on Hilbert space must be Fréchet differentiable in a dense set, say. (Rademacher's theorem shows that Gâteaux differentiability must take place in a dense set, but examples on the real line show that it is too much to demand that the set be a G_δ.) This problem was solved beautifully by D. Preiss [Pr], as stated in Theorem 4.21.

Although Mazur's theorem is both generalized and proved again later in these notes, the original proof (or a reasonable facsimile thereof, since Mazur was essentially looking at Minkowski functionals) was well worth covering. This seemingly magical fact (that continuous convex functions on a separable Banach space must be Gâteaux differentiable on a big set) is what originally stimulated our interest in the subject.

Section 2

2. Monotone operators, subdifferentials and Asplund spaces

Definition 2.1. A set-valued map T from a Banach space E into the subsets of its dual E^* is said to be a *monotone operator* provided

$$\langle x^* - y^*, x - y \rangle \geq 0$$

whenever $x, y \in E$ and $x^* \in T(x)$, $y^* \in T(y)$. We do not require that $T(x)$ be nonempty. The *domain* (or *effective domain*) $D(T)$ of T is the set of all $x \in E$ such that $T(x)$ *is* nonempty.

2.2. Examples.

(a) If f is a continuous convex function on a nonempty open convex subset D of E, then $T(x) = \partial f(x)$ $(x \in D)$, $T(x) = \emptyset$ $(x \in E \setminus D)$ is a monotone operator with $D(T) = D$.

Proof. If $x^* \in \partial f(x)$, $y^* \in \partial f(y)$, then

$$\langle x^*, y - x \rangle \leq f(y) - f(x) \text{ and } -\langle y^*, y - x \rangle = \langle y^*, x - y \rangle \leq f(x) - f(y);$$

now add these two inequalities.

[We will see later that the subdifferential map is a rather special monotone operator. In particular, it has certain continuity properties.]

(b) If H is a real Hilbert space and $T: H \to H$ is a linear map, then T is monotone if and only if it is a positive operator: $\langle Tx, x \rangle \geq 0$ for all x.

(c) A function $\varphi: \mathrm{R} \to \mathrm{R}$ is monotone nondecreasing in the usual sense if and only if it is monotone in the above sense: That is, $\varphi(t_1) \leq \varphi(t_2)$ whenever $t_1 < t_2$ if and only if

$$[\varphi(t_2) - \varphi(t_1)] \cdot (t_2 - t_1) \geq 0 \text{ for all } t_1, t_2 \in R.$$

(d) Monotone *derivatives* necessarily arise from convex functions; more precisely, if f is a real-valued Gâteaux differentiable function on E such that df is monotone, then f is convex.

Proof. Given $x, y \in E$ define, for any $\lambda \in \mathrm{R}$,

$$\phi(\lambda) = f(\lambda x + (1-\lambda)y) - \lambda f(x) - (1-\lambda)f(y).$$

This function is differentiable for all λ and $\phi(0) = 0 = \phi(1)$; we want to show that $\phi \leq 0$ in the interval $[0,1]$. If it were not, there would exist $0 < \lambda_0 < 1$ such that $\phi(\lambda_0) > 0$ and ϕ takes its maximum value in $[0,1]$ at λ_0, hence $\phi'(\lambda_0) = 0$. Suppose $\lambda \in (\lambda_0, 1]$; then

$$\phi'(\lambda) - \phi'(\lambda_0) =$$

$$= \langle df(\lambda x + (1-\lambda y), x - y \rangle - f(x) + f(y) - \langle df(\lambda_0 x + (1-\lambda_0 y), x - y \rangle + f(x) - f(y)$$

$$= \langle df(\lambda x + (1-\lambda)y) - df(\lambda_0 x + (1-\lambda_0)y), x - y \rangle.$$

Now, $x - y = \frac{1}{(\lambda - \lambda_0)}\{[\lambda x + (1-\lambda)y] - [\lambda_0 x + (1-\lambda_0)y]\}$, so by monotonicity of df (and the fact that $\frac{1}{(\lambda - \lambda_0)} > 0$), we have $\phi'(\lambda) \geq \phi'(\lambda_0) = 0$, that is, ϕ is nondecreasing on $(\lambda_0, 1]$, a contradiction.

(e) This example arises in fixed-point theory. Let C be a bounded closed convex nonempty subset of Hilbert space H and let U be a (generally nonlinear) nonexpansive map of C into itself: $\|U(x) - U(y)\| \leq \|x - y\|$ for all $x, y \in C$. Let I denote the identity map in H; then $T = I - U$ is monotone, with $D(T) = C$. Indeed, for all x, $y \in C$,

$$\langle T(x) - T(y), x - y \rangle = \langle x - y - (U(x) - U(y)), x - y \rangle =$$

$$= \|x - y\|^2 - \langle U(x) - U(y), x - y \rangle \geq \|x - y\|^2 - \|U(x) - U(y)\| \cdot \|x - y\| \geq 0.$$

Note that 0 is in the range of T if and only if U has a fixed point in C; this hints at the importance for applications of studying the *ranges* of monotone operators (something we will not do, but which is done, for instance, in [Au-Ek], [Pa-Sb] and [Ze]).

(f) Again, in Hilbert space, let C be a nonempty closed convex set and let P be the metric projection of H onto C defined in Example 1.14 (d). Since (as shown there) P is the Fréchet derivative of a convex continuous function on H, it is monotone. More directly, this is an immediate consequence of inequality (1.6).

Definition 2.3. Let X, Y be Hausdorff topological spaces and suppose that $T: X \to 2^Y$ is a set-valued mapping from X into the subsets of Y. If A is a subset of E, we define $T(A) = \cup\{T(x): x \in A\}$. We say that T is *upper semicontinuous at the point* $x \in X$ if, for each open set V in Y containing $T(x)$, there is an open neighborhood U of x such that $T(U) \subset V$. Upper semicontinuity on a *set* is defined in the obvious way.

2.4. Exercises.

(a) Prove that a set-valued map $T: E \to 2^{E^*}$ is norm-to-weak* [norm-to-norm] upper semicontinuous at $x \in E$ if and only if for every weak* open set [norm open set] V containing $T(x)$ and every $\{x_n\} \subset D(T)$ with $\|x_n - x\| \to 0$, we have $T(x_n) \subset V$ for all sufficiently large n. (Equivalently, $T[B(x; \delta)] \subset V$ for all sufficiently small $\delta > 0$.)

(b) Assuming that $T(x)$ is a single point, prove that T is norm-to-norm upper semicontinuous at x if and only if

$$\lim_{\delta \to 0+} \operatorname{diam} T[B(x; \delta)] = 0.$$

Proposition 2.5. *If f is a continuous convex function on the open convex subset D of E, then the subdifferential map $x \to \partial f(x)$ is norm-to-weak* upper semicontinuous on D.*

Proof. We must show that if $x \in D$ and W is any weak* open subset of E^* containing $\partial f(x)$, then for any sequence $\{x_n\} \subset D$ with $x_n \to x$, we have $\partial f(x_n) \subset W$ for all sufficiently large n. If not, then there exists a subsequence (call it $\{x_n\}$) and $x_n^* \in \partial f(x_n) \backslash W$. By local boundedness of the subdifferential map, we can assume that there is a (weak* compact) closed ball which contains the sets $\partial f(x_n)$ for all sufficiently large n. Let x^* be a weak* cluster point of the sequence $\{x_n^*\}$; it is easily verified that x^* is in $\partial f(x) \backslash W$, a contradiction.

Lemma 2.6. *Suppose that f is continuous and convex on a nonempty open convex subset D of E and that it is Fréchet differentiable at a point x in D. Then the subdifferential map ∂f is norm-to-norm upper semicontinuous at x.*

Proof. We want to show that given any norm open neighborhood V of the functional $x^* = f'(x)$ there exists a neighborhood of x which is mapped into V by ∂f. If this were to fail, we could choose $\epsilon > 0$, a sequence of points $\{x_n\} \subset D$ and $x_n^* \in \partial f(x_n)$ for each n such that $\|x_n - x\| \to 0$ while $\|x_n^* - x^*\| > 2\epsilon$. Consequently, there would exist $z_n \in E$, $\|z_n\| = 1$, such that $\langle x_n^* - x^*, z_n \rangle > 2\epsilon$. By Fréchet differentiability of f at x there would exist $\delta > 0$ such that $x + y \in D$ and

$$f(x + y) - f(x) - \langle x^*, y \rangle \le \epsilon \|y\|$$

whenever $\|y\| \le \delta$. Since $x_n^* \in \partial f(x_n)$, we have

$$\langle x_n^*, (x + y) - x_n \rangle \le f(x + y) - f(x_n) \text{ so}$$

$$\langle x_n^*, y \rangle \le f(x + y) - f(x) + \langle x_n^*, x_n - x \rangle + f(x) - f(x_n)$$

whenever $\|y\| \le \delta$. Let $y_n = \delta z_n$, so $\|y_n\| = \delta$ and

$$2\epsilon\delta < \langle x_n^* - x^*, y_n \rangle \le$$

$$\le [f(x + y_n) - f(x) - \langle x^*, y_n \rangle] + \langle x_n^*, x_n - x \rangle + f(x) - f(x_n) \le$$

$$\le \epsilon\delta + \langle x_n^*, x_n - x \rangle + f(x) - f(x_n).$$

Since ∂f is locally bounded and $|\langle x_n^*, x_n - x \rangle| \le \|x_n^*\| \cdot \|x_n - x\|$, this term converges to 0, while $f(x) - f(x_n) \to 0$ since f is continuous. But this would yield $2\epsilon\delta \le \epsilon\delta$, an impossibility which completes the proof.

Definition 2.7. A *selection* φ for a set-valued map T is a single-valued mapping satisfying $\varphi(x) \in T(x)$ for each $x \in D(T)$.

Proposition 2.8. *If f is convex and continuous on the nonempty open convex subset D of E, then it is Gâteaux [Fréchet] differentiable at a point $x \in D$ if and only if there is a selection φ for the subdifferential map ∂f which is norm-to-weak* [norm-to-norm] continuous at x.*

Proof. Suppose that a selection φ for ∂f exists. Since $\varphi(x) \in \partial f(x)$, we must have $\langle \varphi(x), y - x \rangle \leq f(y) - f(x)$ for all $y \in D$. For such y we also have $\varphi(y) \in \partial f(y)$, so $\langle \varphi(y), x - y \rangle \leq f(x) - f(y)$. These inequalities combine to show that for all $y \in D$,

$$0 \leq f(y) - f(x) - \langle \varphi(x), y - x \rangle \leq \langle \varphi(y) - \varphi(x), y - x \rangle. \qquad (2.1)$$

Suppose φ is norm-to-weak* continuous at x. By taking $v \in E$ and $t > 0$ sufficiently small, one can make the substitution $y = x + tv$ in (2.1), divide by t, let $t \to 0^+$ and use the weak* continuity of φ to conclude that $\langle \varphi(x), v \rangle = d^+ f(x)(v)$ for each $v \in E$. That is, $d^+ f(x)$ is linear, hence f is Gâteaux differentiable at x. If φ is norm-to-norm continuous at x, then the fact the last term in (2.1) is dominated by $\|\varphi(y) - \varphi(x)\| \cdot \|y - x\|$ implies that f is Fréchet differentiable at x.

To prove the two converses, suppose that f is Gâteaux differentiable at x, so that $\partial f(x)$ is a singleton. Since ∂f is norm-to-weak* upper semicontinuous, *any* selection for ∂f will necessarily be norm-to-weak* continuous. A similar argument applies if f is Fréchet differentiable at x, since ∂f is norm-to-norm upper semicontinuous, by Lemma 2.6.

An interesting consequence of this result is that Fréchet differentiable convex functions are necessarily C^1.

Corollary. *If f is convex and Fréchet differentiable on the open convex set D, then $x \to f'(x)$ is norm-to-norm continuous in D.*

We are approaching Asplund's theorem that a Banach space E with separable dual is an Asplund space. Since the subdifferential ∂f of a continuous convex function is a monotone map and since Fréchet differentiability of f can be characterized in terms of the norm-to-norm upper semicontinuity of ∂f, we will have proved a generalization of Asplund's theorem once we have proved that monotone maps into a separable dual space are generically single-valued and norm-to-norm upper semicontinuous. We follow an approach devised by D. Preiss and L. Zajíček [Pr-Z], who have shown that the set of points of nondifferentiability is even smaller than merely being a countable union of nowhere dense closed sets. We need a couple of definitions, the first of which has been used in other contexts.

Definition 2.9. (a) If $x^* \in E^*$, $x^* \neq 0$, and $0 < \alpha < 1$, define

$$K(x^*; \alpha) = \{x \in E \colon \alpha \|x\| \cdot \|x^*\| \leq \langle x^*, x \rangle\}.$$

It is easily verified that $K(x^*; \alpha)$ is a closed convex cone (we call it an α-*cone*) and that its nonempty interior is $\{x \in E \colon \alpha \|x\| \cdot \|x^*\| < \langle x^*, x \rangle\}$.

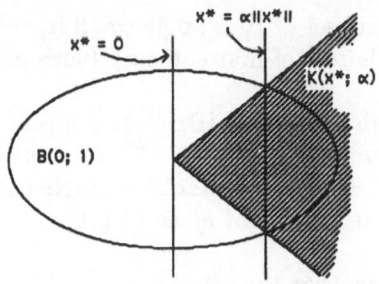

Fig. 2.1

As the sketch indicates, when α becomes closer to 0, then the cone becomes wider.

(b) A subset $M \subset E$ is said to be α-*cone meager* (where $0 < \alpha < 1$) if for every $x \in M$ and $\epsilon > 0$ there exist $z \in B(x; \epsilon)$ and $0 \neq x^* \in E^*$ such that

$$M \cap [z + \text{ int } K(x^*; \alpha)] = \emptyset.$$

(Note that $z + \text{ int } K(x^*; \alpha) = \{y \in E : \alpha \|x^*\| \cdot \|y - z\| < \langle x^*, y - z \rangle \}$.)

To say that M is α-cone meager means that any point of M can be approached arbitrarily closely by vertices of cones whose interiors lie in the complement of M; Figure 2.2 below illustrates the situation (for a set M which is clearly *not* α-cone meager).

(c) The set M is said to be *angle-small* if for every $0 < \alpha < 1$ it can be expressed as a countable union of α-cone meager sets.

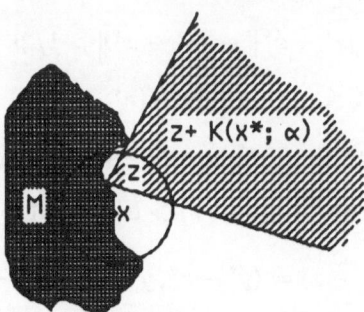

Fig. 2.2

2.10. Exercise.

Prove that if M is α-cone meager for some $\alpha > 0$, then $\overline{M}$ has empty interior, hence show that any angle-small set is of first category.

Being α-cone meager has geometric as well as topological implications. Notice, for instance, that if M is the union in $\mathbf{R}^2$ of the unit circle and the origin, then it is nowhere dense, but it is not α-cone meager for any $\alpha > 0$. (It is, however, the union of two sets, each of which is α-cone meager for every $\alpha > 0$.) Since an α-cone meager subset of $\mathbf{R}$ can contain at most two points, it is easily seen that a subset of $\mathbf{R}$ is angle-small if and only if it is countable. There exist uncountable sets of first category (such as the Cantor set).

Theorem 2.11. (Preiss-Zajiček [Pr-Z]). *Suppose that the Banach space E has separable dual and that $T: E \to 2^{E^*}$ is monotone. Then there exists an angle-small set $A \subset D(T)$ such that T is single-valued and norm-to-norm upper semicontinuous at each point of $D(T) \setminus A$.*

Proof. It suffices to show that the set

$$A = \{x \in D(T): \lim_{\delta \to 0+} \operatorname{diam} T[B(x; \delta)] > 0\}$$

is angle small. We can obviously write $A = \cup A_n$ where

$$A_n = \{x \in D(T): \lim_{\delta \to 0+} \operatorname{diam} T[B(x; \delta)] > 1/n\}.$$

Let $\{x_k^*\}$ be a dense sequence in E^* and suppose $0 < \alpha < 1$. Define

$$A_{n,k} = \{x \in A_n: \operatorname{dist}(x_k^*, T(x)) < \alpha/4n\};$$

we will show that each $A_{n,k}$ is α-cone meager. Suppose that $x \in A_{n,k}$ and that $\epsilon > 0$. Since $x \in A_n$, there exist $0 < \delta < \epsilon$ and $z_1, z_2 \in B(x; \delta)$ and $z_i^* \in T(z_i)$ such that $\|z_1^* - z_2^*\| > 1/n$. Thus, for any $x^* \in T(x)$, one of $\|z_i^* - x^*\| > 1/2n$. Since $\operatorname{dist}(x_k^*, T(x)) < \alpha/4n$, we can choose $x^* \in T(x)$ such that $\|x_k^* - x^*\| < \alpha/4n$ and therefore there exist points $z \in B(x; \epsilon)$ and $z^* \in T(z)$ such that

$$\|z^* - x_k^*\| \geq \|z^* - x^*\| - \|x_k^* - x^*\| > 1/2n - \alpha/4n > 1/4n.$$

We want to show that $A_{n,k} \cap (z + \operatorname{int} K(z^* - x_k^*; \alpha)) = \emptyset$, that is,

$$A_{n,k} \cap \{y \in E : \langle z^* - x_k^*, y - z \rangle > \alpha \|z^* - x_k^*\| \cdot \|y - z\|\} = \emptyset.$$

Now, if $y \in D(T)$ and $\langle z^* - x_k^*, y - z \rangle > \alpha \|z^* - x_k^*\| \cdot \|y - z\|$ and if $y^* \in T(y)$, then

$$\langle y^* - x_k^*, y - z \rangle = \langle y^* - z^*, y - z \rangle + \langle z^* - x_k^*, y - z \rangle \geq$$

$$\geq \langle z^* - x_k^*, y - z \rangle > \alpha \|z^* - x_k^*\| \cdot \|y - z\| \geq (\alpha/4n)\|y - z\|.$$

It follows that $\|y^* - x_k^*\| \geq \alpha/4n$, so y is not in $A_{n,k}$.

This theorem, combined with Exercise 2.10, leads to the following theorem.

Theorem 2.12 (Asplund [Asp]). *If the dual E^* of the Banach space E is separable, then E is an Asplund space.*

Proof. If f is continuous and convex on the open convex set $D \subset E$, then ∂f is monotone, so by Theorem 2.11, it is single-valued and norm-to-norm upper semicontinuous at the points of some dense G_δ subset G of D. Thus, any selection for ∂f is continuous at the points of G, so by Proposition 2.8, f is Fréchet differentiable at the points of G.

2.13. Examples.

Here are some well-known examples of Banach spaces with separable duals: The sequence space c_0 (but not its dual ℓ^1). The spaces ℓ^p and $L^p[0,1]$, provided $1 < p < \infty$. More generally, any separable reflexive Banach space.

We will see later that separable Asplund spaces are *characterized* by the fact that their duals are separable. A more general fact, one of the most beautiful and fundamental results in this area, is that E is an Asplund space if and only if every separable subspace of E has a separable dual. Half of this latter result is a corollary to the next theorem.

Theorem 2.14. *A Banach space E is an Asplund space if every separable closed subspace F of E is an Asplund space.*

Proof. Suppose that f is continuous and convex on the nonempty open convex subset D of E and suppose that the set G of all points $x \in D$ where f is Fréchet differentiable is *not* dense in D; we will produce a separable subspace F of E such that $F \cap D \neq \emptyset$ and the points of Fréchet differentiability of $f|_F$ are not dense in $F \cap D$. For each n define (as in the proof of Prop. 1.25) $G_n(f)$ to be the open set of all $x \in D$ for which there exists $\delta > 0$ satisfying

$$\sup_{\|y\|=1} \frac{f(x + \delta y) + f(x - \delta y) - 2f(x)}{\delta} < 1/n.$$

Since D is a Baire space and since $G = \cap G_n(f)$, we conclude that for some $m > 0$, the set $G_m(f)$ is not dense in D, that is, there exists a nonempty open subset $U \subset D \setminus G_m(f)$. We will next define by induction an increasing sequence $\{F_k\}$ of separable subspaces of E. (The desired subspace F will be the closure of their union.) First, choose $x_1 \in U$. It follows (take $\delta = 1/j$, $j = 1, 2, 3, \ldots$ and use the monotonicity of the difference quotients for f) that there exists a sequence $\{y_{1,j}\}$ with $\|y_{1,j}\| = 1$, such that for all $\delta > 0$,

$$\sup_j \frac{f(x_1 + \delta y_{1,j}) + f(x_1 - \delta y_{1,j}) - 2f(x_1)}{\delta} \geq 1/2m.$$

Let F_1 denote the closed linear span of x_1 and $\{y_{1,j}\}$. Clearly, F_1 is separable and $F_1 \cap U$ is nonempty (since it contains x_1). Given an increasing sequence of separable subspaces $F_1, \ldots, F_k$, define F_{k+1} as follows: Let $\{x_{k,p}\}$, $p = 1, 2, 3, \ldots$ be a countable dense subset of $F_k \cap U$ and for each p choose a sequence $\{y_{p,j}\}$ with $\|y_{p,j}\| = 1$ such that for all $\delta > 0$

$$\sup_j \frac{f(x_{k,p} + \delta y_{p,j}) + f(x_{k,p} - \delta y_{p,j}) - 2f(x_{k,p})}{\delta} \geq 1/2m.$$

Let F_{k+1} be the closed linear span of F_k and $\{x_{k,p}\} \cup \{y_{p,j}\}$, $p, j \geq 1$. Let $F = \overline{\cup F_k}$; then $\{x_{k,p} : k, p = 1, 2, 3, \ldots\}$ is a dense subset of $F \cap U$. It is clear that for each k, p, the point $x_{k,p}$ is not in $G_{2m}(f|_F)$ and – since $U \subset D$ – the sequence $\{x_{k,p}\}$ is in $D \cap U$, that is, $\{x_{k,p}\}$ is a subset of $(D \cap U) \setminus G_{2m}(f|_F)$. Since $G_{2m}(f|_F)$ is open in F, we must have $F \cap U \subset (D \cap U) \setminus G_{2m}(f|_F)$, which implies that $f|_F$ fails to be Fréchet differentiable at each point of the relatively open nonempty subset $F \cap U$ of $F \cap D$.

Corollary 2.15. *A Banach space E is an Asplund space if every separable subspace of E has separable dual.*

It is immediate from this that every reflexive Banach space is an Asplund space.

2.16 Exercise.

Show that for any set Γ, every separable subspace of $c_0(\Gamma)$ has separable dual, so $c_0(\Gamma)$ is an Asplund space. (Recall that $c_0(\Gamma)$ is the supremum-normed space of all functions $x = (x_\gamma)$ on Γ such that for all $\epsilon > 0$, there exists a finite set $\Gamma_\epsilon \subset \Gamma$ such that $|x_\gamma| < \epsilon$ for every γ in $\Gamma \setminus \Gamma_\epsilon$.) Hint: Show that if F is a separable subspace of $c_0(\Gamma)$, then there exists a countable subset $\Gamma_1 \subset \Gamma$ such that F can be identified with a subspace of $c_0(\Gamma_1)$.

Definition 2.17. By a *slice* of a nonempty set A we mean a (necessarily nonempty) subset of A of the form

$$S(x^*, A, \alpha) = \{x \in A : \langle x^*, x \rangle > \sigma_A(x^*) - \alpha\}$$

where $x^* \in E^*$, $\alpha > 0$ and $\sigma_A(x^*) = \sup\{\langle x^*, x \rangle : x \in A\}$.

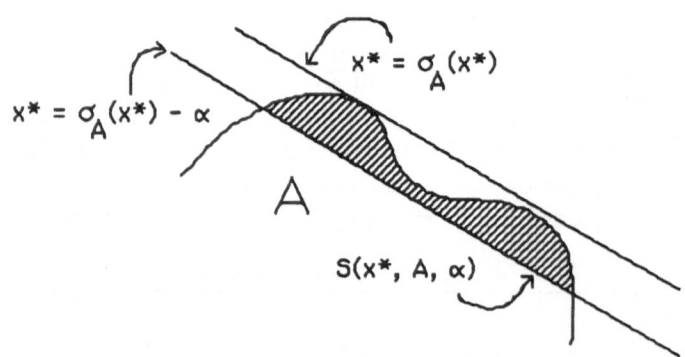

Fig. 2.3

If $A \subset E^*$, then we can also define a *weak* slice* of A to be a slice where the defining linear functional comes from E (rather than from E^{**}). For the moment, in fact, we will be working with weak* slices.

We say that a nonempty subset A *admits slices of arbitrarily small diameter* if, for every $\epsilon > 0$ there exists a slice of A of diameter less than ϵ. (This property is equivalent to a concept called "dentability", so it is sometimes given the same name.)

The following lemma might seem a bit peculiar at first, but – as we will see later – its converse is also valid, so it contains a characterization (a useful one) of Asplund spaces.

Lemma 2.18. *If E is an Asplund space, then every nonempty bounded subset A of E^* admits weak* slices of arbitrarily small diameter.*

Proof. Given A, nonempty and bounded in E^*, define the sublinear functional p on E by $p(x) = \sigma_A(x) \equiv \sup\{\langle x^*, x \rangle : x^* \in A\}$. Since A is bounded, p is necessarily continuous: $p(x) \leq M\|x\|$ if $\|x^*\| \leq M$ for every $x^* \in A$. Suppose that there exists $\epsilon > 0$ such that every weak* slice of A has diameter greater than ϵ; we will show that p is nowhere Fréchet differentiable. Indeed, given any $x \in E$, for each $n \geq 1$ the weak* slice $S(x, A, \epsilon/3n)$ has diameter greater than ϵ. It follows that there exist $x_n^*, y_n^* \in S(x, A, \epsilon/3n)$ with $\|x_n^* - y_n^*\| > \epsilon$, that is

$$x_n^*, y_n^* \in A, \qquad \langle x_n^*, x \rangle > p(x) - \epsilon/3n, \qquad \langle y_n^*, x \rangle > p(x) - \epsilon/3n$$

and $\langle x_n^* - y_n^*, x_n \rangle > \epsilon$ for some $x_n \in E$ with $\|x_n\| = 1$. Thus

$$p[x + (1/n)x_n] + p[x - (1/n)x_n] - 2p(x) \geq$$

$$\geq \langle x_n^*, x + (1/n)x_n \rangle + \langle y_n^*, x - (1/n)x_n \rangle - \langle x_n^* + y_n^*, x \rangle - 2\epsilon/3n =$$

$$= (1/n)\langle x_n^* - y_n^*, x_n \rangle - 2\epsilon/3n > \epsilon/n - 2\epsilon/3n = \epsilon/3n.$$

If we divide through by $1/n$ it becomes evident (using Proposition 1.23) that p is not Fréchet differentiable at x.

We can now characterize separable Asplund spaces.

Theorem 2.19. *A separable Banach space E is an Asplund space if and only if E^* is separable.*

Proof. It was shown in Theorem 2.12 that if E^* is separable, then E is an Asplund space. Suppose then, that E is a separable Asplund space. If E^* is not separable, then its unit ball B^* is not separable, hence there is an uncountable set $A \subset B^*$ and $n \geq 1$ such that $\|x^* - y^*\| > 1/n$ whenever x^*, y^* are distinct points of A. (Look at maximal $1/n$-nets in B^*.) Since E is separable, B^* (in its weak* topology) is compact and metrizable, hence satisfies the second axiom of countability. Thus, A has at most countably many points which are not weak* condensation points; we assume that these points have been deleted from A. Let S be any weak* slice of A. Since such a slice is weak* relatively open and since no point of A is weak* isolated, it must contain two distinct points of A, so its diameter is greater than $1/n$, contradicting Lemma 2.18.

We next look at some additional properties which distinguish subdifferentials within the class of monotone operators, proving some basic results along the way.

Definition 2.20. A set-valued map $T: E \to 2^{E^*}$ is said to be *n-cyclically monotone* provided

$$\sum_{1 \leq k \leq n} \langle x_k^*, x_k - x_{k-1} \rangle \geq 0$$

whenever $n \geq 2$ and $x_0, x_1, x_2, \ldots x_n \in E$, $x_n = x_0$, and $x_k^* \in T(x_k)$, $k = 1, 2, 3, \ldots, n$. We say that T is *cyclically monotone* if it is *n*-cyclically monotone for every n. Clearly, a 2-cyclically monotone operator is monotone.

2.21 Examples.

(a) The linear map in R^2 defined by $T(x_1, x_2) = (x_2, -x_1)$ is positive, hence monotone, but it is not 3-cyclically monotone: Look at the points $(1, 1)$, $(0, 1)$ and $(1, 0)$.

(b) Let f be a continuous convex function on an open convex set; then ∂f is cyclically monotone. (As we will see, this is almost the only example.)

Definition 2.22. A subset G of $E \times E^*$ is said to be *monotone* provided $\langle x^* - y^*, x - y \rangle \geq 0$ whenever $(x, x^*), (y, y^*) \in G$. If $T: E \to 2^{E^*}$ is a monotone operator then its graph

$$G(T) = \{(x, x^*) \in E \times E^* : x^* \in T(x)\}$$

is a monotone set. A monotone set is said to be *maximal monotone* if it is maximal in the family of monotone subsets of $E \times E^*$, ordered by inclusion. We say that a monotone *operator* T is maximal monotone provided its graph is a maximal monotone set.

There is an obvious one-to-one correspondence between monotone sets and monotone operators. An easy application of Zorn's lemma shows that *every monotone operator T can be extended to a maximal monotone operator $\overline{T}$, in the sense that $G(T) \subset G(\overline{T})$.*

2.23 Exercises.

(a) Prove that a monotone operator $T: E \to 2^{E^*}$ is maximal monotone if and only if the following condition holds: Whenever $y \in E$, $y^* \in E^*$ and

$$\langle y^* - x^*, y - x \rangle \geq 0 \text{ for all } x \in D(T), \quad x^* \in T(x)$$

then it necessarily follows that $y^* \in T(y)$.

(b) Prove that if T is maximal monotone, then $T(x)$ is convex set, for every $x \in E$.

(c) Show that the operator in Example 2.21(a) is maximal monotone.

The next theorem is due to G. Minty. A much more general result has been proved by Rockafellar and will be presented in the next section (Theorem 3.24). We start with a simple and useful result.

Proposition 2.24. *If f is convex on D and continuous at $x \in D$, then for all $y \in E$,*
$$d^+ f(x)(y) = \sup\{\langle x^*, y \rangle : x^* \in \partial f(x)\}$$
and this supremum is attained at some point $x^* \in \partial f(x)$.

Proof. As shown in Prop. 1.8, if $x^* \in \partial f(x)$, then $\langle x^*, y \rangle \leq d^+ f(x)(y)$ for all y. On the other hand, by Corollary 1.7, $d^+ f(x)$ is a continuous sublinear functional, so for any $y \neq 0$ we can use the Hahn-Banach theorem to find $x^* \in E^*$ such that $\langle x^*, z \rangle \leq d^+ f(x)(z)$ for all $z \in E$ (so $x^* \in \partial f(x)$ by Proposition 1.8) and $\langle x^*, y \rangle = d^+ f(x)(y)$, which completes the proof.

Theorem 2.25. *If f is continuous and convex on all of E, then its subdifferential map ∂f is maximal monotone.*

Proof. By Exercise 2.23(a), to show that ∂f is maximal it suffices to show that whenever $y \in E$ and $y^* \in E^*$ are such that $y^* \notin \partial f(y)$, then there exist $x \in E$ and $x^* \in \partial f(x)$ such that $\langle y^* - x^*, y - x \rangle < 0$. To simplify the proof we can replace f by the convex continuous function g defined by $g(x) = f(x + y) - \langle y^*, x \rangle$. It is easily verified that one has $x^* \in \partial g(x)$ if and only if $x^* + y^* \in \partial f(x + y)$. Thus, if $y^* \notin \partial f(y)$, then $0 \notin \partial g(0)$ and if there exist x^* and x with $x^* \in \partial g(x)$ such that $\langle x^*, x \rangle < 0$, then for $z = x + y$ and $z^* = x^* + y^*$ we have $z^* \in \partial f(z)$ and $\langle y^* - z^*, y - z \rangle = \langle x^*, x \rangle < 0$. Assume, then, that $y = 0$ and $y^* = 0$; we want to produce $x \in E$ and $x^* \in \partial f(x)$ such that $\langle x^*, x \rangle < 0$. By Prop. 1.25, we know that 0 is not a global minimum for f, so there exists a point $x_1 \in E$ such that $f(0) > f(x_1)$. Consider the convex function $h(t) = f(tx_1), 0 \leq t \leq 1$. Its right hand derivative at a point $t_0 \in (0, 1)$ is clearly equal to $d^+ f(t_0 x_1)(x_1)$. Suppose this quantity were nonnegative for each such t_0; by one form of the mean value theorem (see, for instance, [Fl, p. 22]), this would imply that $h(0) \leq h(1)$, a contradiction. Thus, it is necessarily negative for some $0 < t_0 < 1$, so by homogeneity (and letting $x = t_0 x_1$), we have $d^+ f(x)(x) < 0$. By Prop. 2.24 above, there must exist $x^* \in \partial f(x)$ such that $\langle x^*, x \rangle = d^+ f(x)(x) < 0$, which completes the proof.

2.26 Example.

In a Banach space E define $f(x) = (1/2)\|x\|^2$ and define $J = \partial f$; the maximal monotone map J is called the *duality mapping* for E. Explicitly,
$$J(x) = \{x^* \in E^* : \langle x^*, x \rangle = \|x^*\| \cdot \|x\| \text{ and } \|x^*\| = \|x\|\}.$$

Proof. It is readily computed that $d^+ f(x)(y) = \|x\| \cdot d^+ \|x\|(y)$. If $x = 0$, then $d^+ f(0)(y) = 0$ for all y, hence is linear and therefore $\partial f(0) = \{0\}$. Suppose, then, that $x \neq 0$. We know (by Prop. 1.8) that $x^* \in \partial f(x)$ if and only if $x^* \leq d^+ f(x)$, that is, if and only if $\|x\|^{-1} x^* \leq d^+ \|x\|$, which is equivalent to $y^* \equiv \|x\|^{-1} x^* \in \partial \|x\|$, that is, if and only if $\langle y^*, y - x \rangle \leq \|y\| - \|x\|$ for all $y \in E$. If, in this last inequality, we take $y = x + z, \|z\| \leq 1$ and apply the triangle inequality, we conclude that $\|y^*\| \leq 1$. If we take $y = 0$, we conclude that $\|x\| \leq \langle y^*, x \rangle \leq \|y^*\| \cdot \|x\|$, so $\|y^*\| = 1$ and $\langle y^*, x \rangle = \|x\|$, which is equivalent to what we want to prove. The converse is easy: If $\|y^*\| = 1$ and $\langle y^*, x \rangle = \|x\|$, then for all y in E, we necessarily have $\langle y^*, y - x \rangle \leq \|y\| - \|x\|$, so $y^* \in \partial \|x\|$.

If the norm of E is Gâteaux differentiable, then obviously $J(x)$ contains only one element; by a slight abuse of notation, we denote it also by $J(x)$.

We now know that subdifferential maps of continuous convex functions are maximal cyclically monotone; we will discuss later the fact that maximal cyclically monotone mappings are the subdifferentials of certain convex functions. These (and other) considerations require a fundamental property of monotone operators which was easy to prove for subdifferentials of continuous convex functions (in Prop. 1.11); namely, that they are locally bounded at points in the interior of their domain.

Definition 2.27.

a) A monotone operator $T: E \rightarrow 2^{E^*}$ is *locally bounded at* $x \in D(T)$ provided there exist $M > 0$ and $\delta > 0$ such that $\|y^*\| \leq M$ whenever $y \in B(x; \delta) \cap D(T)$ and $y^* \in T(y)$.

b) A subset (not necessarily convex) $A \subset E$ which contains the origin is said to be *absorbing* if $E = \cup\{\lambda A : \lambda > 0\}$. Equivalently, A is absorbing if for each $x \in E$ there exists $t > 0$ such that $tx \in A$. A point $x \in A$ is called an *absorbing point of* A if the translate $A - x$ is absorbing.

It is obvious that any interior point of a set is an absorbing point. If A_1 is the union of the unit sphere and $\{0\}$, then A_1 is absorbing, even though it has empty interior.

Theorem 2.28. *Suppose that* $T: E \rightarrow 2^{E^*}$ *is monotone and that* $x \in D(T)$. *If* $x \in$ int $D(T)$, *or, more generally, if* x *is an absorbing point of* $D(T)$, *then* T *is locally bounded at* x.

Proof. By choosing any $x^* \in T(x)$ and replacing T by the monotone operator $y \rightarrow T(y + x) - x^*$, we lose no generality in assuming that $x = 0$ and that $0 \in T(0)$. Thus, we want to show that, under these assumptions, T is locally bounded at 0. Define, for $x \in E$,

$$f(x) = \sup\{\langle y^*, x - y \rangle : y \in D(T), \|y\| \leq 1 \text{ and } y^* \in T(y)\}$$

and let $C = \{x \in E : f(x) \leq 1\}$. As the supremum of affine continuous functions, f is convex and lower semicontinuous, and hence C is closed and convex. It also contains the origin: First, since $0 \in T(0)$, we must have $f \geq 0$. Second, whenever $y \in D(T)$ and $y^* \in T(y)$, monotonicity implies that $0 \leq \langle y^* - 0, y - 0 \rangle$, so $f(0) \leq 0$. We claim that the closed convex symmetric set $A = C \cap (-C)$ is absorbing, hence, by a standard consequence of the Baire category theorem, is a neighborhood of the origin. It suffices to prove that C is absorbing, so suppose $x \in E$. By hypothesis, $D(T)$ is absorbing so there exists $t > 0$ such that $T(tx) \neq \emptyset$. Choose any element $u^* \in T(tx)$. If $y \in D(T)$ and $y^* \in T(y)$, then by monotonicity

$$\langle y^*, tx - y \rangle \leq \langle u^*, tx - y \rangle.$$

Consequently,

$$f(tx) \leq \sup\{\langle u^*, tx - y \rangle : y \in D(T), \|y\| \leq 1\} < \langle u^*, tx \rangle + \|u^*\| < \infty.$$

Choose $0 < \lambda < 1$ such that $\lambda f(tx) < 1$. By convexity

$$f(\lambda tx) \leq \lambda f(tx) + (1 - \lambda)f(0) = \lambda f(tx) < 1,$$

so $\lambda tx \in C$. Thus, A is a neighborhood of 0 and hence there exists $\delta > 0$ such that $f(x) \leq 1$ whenever $\|x\| \leq 2\delta$. Equivalently, if $\|x\| \leq 2\delta$, then $\langle y^*, x \rangle \leq \langle y^*, y \rangle + 1$ whenever $y \in D(T)$, $\|y\| \leq 1$ and $y^* \in T(y)$. Thus, if $y \in B(0; \delta) \cap D(T)$ and $y^* \in T(y)$, then

$$2\delta\|y^*\| = \sup\{\langle y^*, x \rangle : \|x\| \leq 2\delta\} \leq \|y^*\| \cdot \|y\| + 1 \leq \delta\|y^*\| + 1,$$

so $\|y^*\| \leq 1/\delta$.

Note that the foregoing result does not require that $D(T)$ be convex. There are trivial examples which show that 0 can be an absorbing point of $D(T)$ but not an interior point (for instance, let T be the restriction of the subdifferential of the norm to the set A_1 defined above). Even if $D(T)$ *is* convex and T is maximal monotone, $D(T)$ can have empty interior, as shown by the following example. (In this example, T is an unbounded linear operator, hence it is not locally bounded at any point and therefore $D(T)$ has no absorbing points.)

Example.

In the Hilbert space ℓ^2, let $D = \{x = (x_n) \in \ell^2 : (2^n x_n) \in \ell^2\}$ and define $Tx = (2^n x_n)$, $x \in D$. Then $D(T) = D$ is a proper dense linear subspace of ℓ^2 and T is a positive operator, hence is monotone. We use Exercise 2.23 (a) to show that it is maximal monotone. Suppose, then, that y and y^* are in ℓ^2 and that for all $x \in D(T)$

$$0 \leq \langle y^* - Tx, y - x \rangle \equiv \langle y^*, y \rangle - \langle Tx, y \rangle - \langle y^*, x \rangle + \langle Tx, x \rangle; \qquad (2.2)$$

we want to show that $y \in D(T)$ and that $Ty = y^*$. Fix $n \geq 1$, $m \geq 1$ and $\alpha \in \mathbf{R}$ and let

$$x = (y_1, y_2, \ldots, y_{n-1}, \alpha, y_{n+1}, \ldots, y_{n+m}, 0, 0, \ldots).$$

Since $x \in D(T)$, we can expand the right side of (2.2) and cancel a number of terms to obtain

$$0 \leq \langle y^*, y \rangle - \alpha 2^n y_n - \Sigma_1^{n+m} y_k^* y_k - \alpha y_n^* + y_n^* y_n + 2^n \alpha^2.$$

Letting $m \to \infty$ yields $\alpha 2^n(y_n - \alpha) \leq y_n^*(y_n - \alpha)$ for all $n \geq 1$. Since this is true for arbitrary $\alpha \in \mathbf{R}$, it follows that $y_n^* = 2^n y_n$ for each n. Since $y^* \in \ell^2$ and $y^* = (2^n y_n)$, we conclude that $y \in D(T)$ and $y^* = T(y)$.

It is conceivable that for a maximal monotone T, any absorbing point of $D(T)$ is actually an interior point.

A word of caution is in order at this point. Our knowledge of the structure of the domain of monotone operators is incomplete, although some things are known about $D(T)$ when T is *maximal* monotone. Rockafellar [Ro$_2$] gives two special conditions under which int $D(T) \neq \emptyset$ for a maximal monotone T; in particular, this is the case if the convex hull of $D(T)$ is assumed to have nonempty interior. Under this hypothesis, incidentally, the interior of $D(T)$ is convex, and for any point $x \in D(T)\backslash\text{int } D(T)$, the set $T(x)$ is unbounded.

(Indeed, by the separation theorem there exists $y^* \neq 0$ such that $\langle y^*, x \rangle \geq \langle y^*, u \rangle$ for every point $u \in D(T)$. Thus, for every $\lambda > 0$ and any $x^* \in T(x)$, $u^* \in T(u)$, we have

$$\langle (x^* + \lambda y^*) - u^*, x - u \rangle = \langle x^* - u^*, x - u \rangle + \lambda \langle y^*, x - u \rangle \geq 0,$$

by monotonicity. By maximality, $x^* + \lambda y^* \in T(x)$ for every $\lambda > 0$.)

2.29. Exercises.

(a) Prove that if T is maximal monotone, then $T(x)$ is weak* compact and convex, for all $x \in$ int $D(T)$.

(b) Prove that if T is maximal monotone, then it is norm-to-weak* upper semicontinuous on int $D(T)$.

The following theorem is due to P. Kenderov [Ke$_2$]. Note that by Lemma 2.18, *the "small slices" hypothesis on E^* is satisfied if E is an Asplund space.*

Theorem 2.30. *Suppose that every nonempty bounded subset of the Banach space E^* admits weak* slices of arbitrarily small diameter; then for every maximal monotone operator $T: E \to 2^{E^*}$ there is a dense G_δ subset G of $W =$ int $D(T)$ (which we assume nonempty) such that T is single-valued and norm-to-norm upper semicontinuous at each point of G.*

Proof. For each $n \geq 1$ let G_n be the set of all $x \in W$ for which there is a neighborhood V of x in W such that diam $T(V) < 1/n$. Clearly, T is single-valued and norm-to-norm upper semicontinuous at each point of the intersection $G = \cap G_n$. Since W is a Baire space, we need only show that each of the sets G_n is open and dense in W. They are open by their definition, so it remains to show that each G_n is dense. Let $x \in W$ and let U be any neighborhood of x in W. From Theorem 2.28 we know that T is locally bounded in W, so without loss of generality we can assume that $T(U)$ is a bounded subset of E^*. By hypothesis, there exist $z \in E$ and $\alpha > 0$ such that the weak* slice

$$S \equiv S(z, T(U), \alpha) = \{x^* \in T(U): \langle x^*, z \rangle > \sigma_{T(U)}(z) - \alpha\}$$

has diameter less than $1/n$. If $x^* \in S$, then $x^* \in T(x_1)$ for some point $x_1 \in U$ and $x_0 \equiv x_1 + rz$ is in U for sufficiently small $r > 0$. We claim that $T(x_0) \subset S$. Indeed, if $y^* \in T(x_0)$, then we have

$$0 \leq \langle y^* - x^*, x_0 - x_1 \rangle = r\langle y^* - x^*, z \rangle,$$

which implies that $y^* \in S$. Since $\{x^* \in E^*: \langle x^*, z \rangle > \sigma_{T(U)}(z) - \alpha\}$ is weak* open and since (by Exercise 2.29(b)) T is norm-to-weak* upper semicontinuous, there exists $\delta > 0$ such that $B(x_0; \delta) \subset U$ and $T(y) \subset S$ for any point

$y \in B(x_0; \delta)$; it follows that $T[B(x_0; \delta)]$ has diameter less than $1/n$. This says that $x_0 \in G_n \cap U$, which completes the proof.

We can now prove the converse to Lemma 2.18, obtaining a characterization of Asplund spaces. We first prove a local extension result for convex functions, proved originally by Asplund (although the proof given below is simpler).

Lemma 2.31. *Suppose that f is continuous and convex on the open convex set $D \subset E$ and that $x_0 \in D$. Then there exists a neighborhood U of x_0 in D and a convex Lipschitzian function $\widetilde{f}$ on E such that $\widetilde{f} = f$ in U.*

Proof. Given $n \geq 1$, $x \in E$, define f_n to be the "inf-convolution" of f and $n\|\cdot\|$:

$$f_n(x) = \inf\{f(y) + n\|x - y\| : y \in D\}.$$

We need to know that $f_n > -\infty$ (at least for all sufficiently large n). To this end, choose $x^* \in \partial f(x_0)$. If $n \geq \|x^*\|$, then for any $x \in E$ and $y \in D$ we have

$$f(y) - f(x_0) \geq \langle x^*, y - x_0 \rangle \geq -n\|y - x_0\| \geq -n\|y - x\| - n\|x - x_0\|,$$

so $f_n(x) \geq f(x_0) - n\|x - x_0\|$. Thus, we will assume below that n is large enough that $f_n > -\infty$. It follows easily from the definition that for $x_1, x_2 \in E$, $0 \leq t \leq 1$ and any $\epsilon > 0$, one has $tf_n(x_1) + (1-t)f_n(x_2) \geq f_n(tx_1 + (1-t)x_2) - \epsilon$, so f_n is convex. Also, $f_n(x) \leq f(x)$ for all $x \in D$ (take $y = x$ in the definition). Moreover, given $u, v \in E$ and $\epsilon > 0$ we can choose $y \in D$ such that

$$f_n(u) > f(y) + n\|u - y\| - \epsilon.$$

Since

$$f_n(v) \leq f(y) + n\|v - y\|$$

we have

$$f_n(v) - f_n(u) < f(y) + n\|v - y\| - [f(y) + n\|u - y\| - \epsilon] \leq n\|v - u\| + \epsilon,$$

which shows that $f_n(u) - f_n(v) \leq n\|u - v\|$ for all u, v. Interchanging u, v proves that f_n has Lipschitz constant n. Since ∂f is locally bounded, there exists a neighborhood U of x_0 and an integer $n > 0$ such that $\partial f(U)$ is contained in the ball nB^*. Suppose that $x \in U$ and choose any functional $x^* \in \partial f(x)$; then $\|x^*\| \leq n$ and for all $y \in D$

$$f(x) \leq f(y) + \langle x^*, x - y \rangle \leq f(y) + n\|x - y\|,$$

which implies that $f(x) \leq f_n(x)$; that is, $f = f_n$ in U.

Theorem 2.32. *A Banach space E is an Asplund space if and only if every nonempty bounded subset of E^* admits weak* slices of arbitrarily small diameter.*

Proof. The necessity portion has already been proved in Lemma 2.18. Suppose, then, that f is continuous and convex on the nonempty open convex set $D \subset E$; we need only show that f' exists on a dense subset of D. Given $x_0 \in D$ and a neighborhood U of x_0, we can assume that U is sufficiently small that the conclusion to Lemma 2.31 holds in U; that is, there exists a convex Lipschitz continuous function $\tilde{f}$ on E such that $\tilde{f} = f$ on U. By Theorems 2.25 and 2.30, there is a dense G_δ subset G of points in E at each of which $\partial \tilde{f}$ is single-valued and norm-to-norm upper semicontinuous, hence any selection for $\partial \tilde{f}$ is norm-to-norm continuous. By Prop. 2.8, $\tilde{f}$ is Fréchet differentiable at each such point. But this implies that f has points of differentiability in U.

As we will see shortly, this characterization has a number of applications. For instance, it is not obvious that a closed subspace M of an Asplund space E is itself an Asplund space; even if we could extend a continuous convex function on an open convex subset of M to one on an open convex subset of E, the (dense) set of points of differentiability of the extension could fail to intersect the (nowhere dense) set M. The Asplund property *is* inherited by closed subspaces, however, and Theorem 2.32 can be used to prove it.

Proposition 2.33. *A closed subspace M of an Asplund space E is itself an Asplund space.*

Proof. We need only show that any bounded nonempty subset A of $M^* = E^*/M^\perp$ admits weak* slices of arbitrarily small diameter. Without loss of generality, we can assume that A is weak* compact and convex. (Any weak* slice of the weak* closed convex hull of A is also a slice of A.) The quotient map $Q: E^* \to M^*$ is of norm one, onto and weak*-to-weak* continuous. Suppose that $\epsilon > 0$. Let B^* be the unit ball of E^*; since Q is an open map, the set $Q(B^*)$ contains a neighborhood of the origin in M^*. Since A is bounded, this implies that there exists $\lambda > 0$ such that $Q(\lambda B^*) = \lambda Q(B^*) \supset A$. Let $C = \lambda B^* \cap Q^{-1}(A)$; clearly, C is weak* compact, convex and $Q(C) = A$. By Zorn's lemma there exists a minimal (under inclusion) set C with these properties; let C_1 be such a minimal set. By hypothesis, there exists a weak* slice $S \equiv S(x, C_1, \alpha)$ of C_1 of diameter less that ϵ; since S is relatively weak* open, the set $A_1 = Q(C_1 \setminus S)$ is a weak* compact convex set which, by the minimality of C_1, is properly contained in A. If $x_1^*, x_2^* \in A \setminus A_1$, there exist $y_1^*, y_2^* \in S$ such that $Q(y_i^*) = x_i^*$ and

$$\|x_1^* - x_2^*\| = \|Q(y_1^* - y_2^*)\| \leq \|y_1^* - y_2^*\| < \epsilon.$$

Thus, $\mathrm{diam}(A \setminus A_1) \leq \epsilon$. By the separation theorem, there exists a weak* slice of A which misses A_1, and hence has diameter at most ϵ. By Theorem 2.32, again, we conclude that M is an Asplund space.

Theorem 2.34. *A Banach space is an Asplund space if and only if every separable closed subspace of E has a separable dual.*

Proof. The sufficiency half of this theorem is precisely Corollary 2.15. The necessity follows from the previous proposition and Theorem 2.19.

Problem. It is a difficult open question whether a closed subspace of a weak Asplund space is itself a weak Asplund space.

At this point we should state the obvious: *The property of being an Asplund (or weak Asplund) space is preserved under linear isomorphisms;* that is, replacing the norm in a Banach space by an equivalent norm will have no effect on the differentiability of functions on the space nor on a topological property like being a dense G_δ. This fact has played an important role in the subject, since many results, starting with Asplund's own theorems, were obtained by showing first that particular classes of spaces admit norms which themselves have good differentiability properties, then using that fact to deduce differentiability for arbitrary continuous convex functions. The close connection between differentiable norms and Asplund spaces is illustrated by the following corollary to Theorem 2.32.

Corollary 2.35. *If a Banach space E is not an Asplund space, then there exists an equivalent norm on E which is nowhere Fréchet differentiable.*

Proof. If E is not an Asplund space, then by Theorem 2.32 there exists a bounded nonempty subset A of E^* and $\epsilon > 0$ such that every weak* slice of A has diameter greater than ϵ. We will use A to construct a bounded symmetric convex set U^* with nonempty interior in E^* such that every weak* closed slice of U^* has diameter greater than ϵ. From the proof of Lemma 2.18, it will follow that the continuous support function p for U^* is nowhere Fréchet differentiable. Since U^* has nonempty interior, p defines an equivalent norm on E. To construct U^*, note that both the convex hull C of $A \cup -A$ and the sum $U^* = C + B^*$, where B^* is the unit ball of E^*, are bounded, symmetric and have the property that all their weak* slices have diameter greater than ϵ. (This last assertion for U^* makes use of the fact that $\sigma_{C+B^*} = \sigma_C + \sigma_{B^*}$.) The set U^* has nonempty interior, since B^* does.

For one of his Gâteaux differentiability theorems, Asplund used the hypothesis that E could be given an equivalent norm which was itself Gâteaux differentiable in a certain strong sense. We will approach this result through maximal monotone operators and another theorem of Kenderov.

Definition 2.36.

(a) A norm on a Banach space E is said to be *strictly convex* (or *rotund*) provided there are no line segments in the unit sphere; equivalently, provided

$$\|x\| = 1 = \|y\| \text{ and } x \neq y \text{ imply } \|\lambda x + (1-\lambda)y\| < \lambda\|x\| + (1-\lambda)\|y\|$$

whenever $0 < \lambda < 1$.

(b) A norm on E is said to be *smooth* provided that for each $x \in E$ with $\|x\| = 1$, there exists a unique element $x^* \in E^*$ such that $\langle x^*, x \rangle = 1$ and $\|x^*\| = 1$. This is the same, clearly, as saying that the duality map $J(x)$ (Example 2.26) is a single point for every $x \neq 0$, which in turn is the same as saying that the norm is Gâteaux differentiable at every nonzero point of E.

It is common terminology to refer to a *space* as being smooth or strictly convex if its norm has that property.

2.37 Exercises.

(a) Prove that if the norm in E is such that its dual norm in E^* is strictly convex [smooth], then it is itself smooth [strictly convex].

(b) Prove that the norm in E is strictly convex if and only if every convex subset of E has at most one point of least norm.

(c) Show that the norm in Hilbert space is both smooth and strictly convex, but that the norms in c_0 and ℓ^1 are neither.

(d) Prove that the norm in E is strictly convex if and only if $\|x + y\| < \|x\| + \|y\|$ whenever x and y are linearly independent.

It is known [Kl], [Tr] that there exist smooth spaces whose duals are not strictly convex; we will see in Section 4 that there also exist strictly convex spaces whose duals are not smooth.

The proof of the next theorem uses the following classical fact: For any lower semicontinuous real-valued function on an open subset of a complete metric space there exists a dense G_δ set of points at which the function is continuous. (See, for instance, [Ch, p.111].)

Theorem 2.38 ([Ke₁]). *Suppose that E admits an equivalent norm whose dual norm is strictly convex and suppose that $T: E \to 2^{E^*}$ is a maximal monotone map such that $W = \operatorname{int} D(T)$ is nonempty. Then there exists a dense G_δ subset of W at every point of which T is single-valued.*

Proof. Assuming that the norm in E^* is strictly convex, we know that for each $x \in W$, the nonempty weak* compact set $T(x)$ has at most one point of least norm – in fact – *exactly* one point of least norm, since dual norms are weak* lower semicontinuous. Define the real-valued function β by

$$\beta(x) = \min\{\|x^*\|: x^* \in T(x)\}, \quad x \in W.$$

We first show that β is lower semicontinuous on W; that is, for any real number λ, the set $\{x \in W : \beta(x) > \lambda\}$ is open. Indeed, since the norm in E^* is weak* lower semicontinuous, the set $V = \{x^* \in E^*: \|x^*\| > \lambda\}$ is weak* open. From Exercise 2.29(b) we know that maximal monotone operators are norm-to-weak* upper semicontinuous. Thus, if $\beta(x) > \lambda$, then $T(x) \subset V$ and hence there exists a neighborhood U of x in W such that $T(U) \subset V$, which implies that $\beta(y) > \lambda$ for all $y \in U$.

We next show that *if β is continuous at $x \in W$, then $T(x)$ consists of a single point.* Indeed, suppose that $\beta(x) = \|x^*\|$, $x^* \in T(x)$, and that there exists $y^* \in T(x)$, with $y^* \neq x^*$; necessarily, $\|y^*\| > \|x^*\|$. We can choose $y \in E$, $\|y\| = 1$, such that $\langle y^*, y \rangle > \|x^*\| = \beta(x)$. By continuity of β at x, there exists $\delta > 0$ such that

$$B(x;\delta) \subset W \text{ and } \beta(z) < \langle y^*, y \rangle \text{ whenever } z \in B(x;\delta).$$

Let $z = x + \delta y \in B(x;\delta)$ and let $z^* \in T(z)$ be such that $\beta(z) = \|z^*\|$. By monotonicity,

$$0 \leq \langle z^* - y^*, z - x \rangle = \delta \langle z^* - y^*, y \rangle,$$

so $\langle z^*, y \rangle \geq \langle y^*, y \rangle > \beta(z) = \|z^*\|$, which is a contradiction. Finally, we complete the proof by applying the general fact about lower semicontinuous functions mentioned above.

Corollary 2.39. *(Asplund) If E admits an equivalent norm which has strictly convex dual norm, then E is a weak Asplund space.*

Proof. The proof is similar to that of Theorem 2.32.

2.40. Exercises.

(a) Suppose that $|\cdot\cdot|$ is an equivalent norm on E^*, (that is, $m \cdot \|x^*\| \leq |x^*| \leq M \cdot \|x^*\|$ for all x^*, where $0 < m \leq M$). Prove that $|\cdot\cdot|$ is the dual of an equivalent norm on E if and only if $|\cdot\cdot|$ is weak* lower semicontinuous.

(b) Suppose that E is separable, that $\{x_n\}$ is a dense sequence in the unit sphere of E and, for $x^* \in E^*$, define

$$L(x^*) = (2^{-n} \langle x^*, x_n \rangle) \in \ell^2.$$

Prove that $|x^*| = \|x^*\| + \|L(x^*)\|_2$ defines an equivalent strictly convex dual norm on E^*. Hence deduce Mazur's theorem from Cor. 2.39.

We next look at a property, weaker than separability, which has implications for both Asplund spaces and weak Asplund spaces.

Definition 2.41. A Banach space E is said to be *weakly compactly generated* (WCG) provided there exists a weakly compact subset K of E whose linear span is dense in E. Since the closed convex hull of a weakly compact subset of a Banach space is weakly compact, one can always assume that K is convex. For background (and proofs of some of the assertions which follow) see J. Diestel's lecture notes [Di].

2.42. Examples.

(a) If E is separable or reflexive, then it is WCG. In the first case, let $\{x_n\}$ be dense in E and let $K = \{n^{-1}x_n/\|x_n\|\} \cup \{0\}$; this is actually compact. In the second case, let K be the unit ball.

(b) The Banach space $c_0(\Gamma)$, for any set Γ, is WCG: If, for each $\gamma \in \Gamma$, we let e_γ denote the usual basis vector, then $\{e_\gamma : \gamma \in \Gamma\} \cup \{0\}$ is weakly compact.

(c) The Banach space $\ell^1(\Gamma)$ is WCG (if and) only if it is separable, that is, if and only if Γ is countable (since weakly compact subsets of $\ell^1(\Gamma)$ are norm compact).

(d) If μ is finite, then $L^1(\mu)$ is WCG. (Since $L^\infty(\mu) \subset L^1(\mu)$ and the inclusion mapping is weak* to weak continuous, the unit ball of $L^\infty(\mu)$ is weakly compact in $L^1(\mu)$.) The ℓ^1-product of a countable family of WCG spaces is WCG, hence $L^1(\mu)$ is WCG if (and only if) μ is σ-finite.

(e) If E is WCG and there exists a bounded linear operator on E having dense range in the Banach space F, then F is WCG.

The next theorem obviously generalizes Theorem 2.12 (that if E^* is separable, then E is an Asplund space).

Theorem 2.43. *If E^* is WCG, then E is an Asplund space.*

Proof. We will apply Cor. 2.15, by showing that for every separable subspace F of E, its dual F^* is separable. Since the latter is isometric to the quotient space $E^*/F^\perp$, and since a continuous linear image of a WCG space is WCG, we can assume that F^* is WCG. Let K be a weakly compact subset of F^* whose linear span is dense in F^*; it clearly suffices to prove that K is norm separable. But this is immediate from the following general lemma.

Lemma 2.44. *If E is a separable Banach space and if K is a weakly compact subset of E^*, then K is norm separable.*

Proof. Since K is weakly compact, it is also weak* compact and (K, w) and (K, w^*) are homeomorphic. Any weak* compact subset of the dual of a separable space is weak* metrizable, so (K, w^*) is separable, and hence (K, w) is separable. Let A be a countable weakly dense subset of K and let L be the set of all rational linear combinations of elements of A, so L is countable and hence its norm closure $\overline{L}$ is norm separable. Clearly, $\overline{L}$ contains *all* linear combinations of members of A, hence $\overline{L}$ is the smallest norm closed linear subspace containing A. Since $\overline{L}$ is weakly closed, it contains the weak closure of A, that is, it contains K. Thus, K is norm separable, since it is a subset of the norm separable space $\overline{L}$.

An argument (due to P. D. Morris) which uses some of the same ideas as above can be used to prove Kuo's theorem that if E^* *is isomorphic to a subspace of a WCG space, then E is an Asplund space*. This may be found in R. Bourgin's lecture notes [Bou]; since H. P. Rosenthal has shown (see [Di]) that there exist subspaces of the WCG space $L^1[\mu]$, μ finite, which are dual spaces but are not themselves WCG, this is a generalization of Theorem 2.43.

The converse to Theorem 2.43 is false: By Exercise 2.16, $c_0(\Gamma)$ is an Asplund space for every set Γ, while its dual $\ell^1(\Gamma)$ is not WCG if Γ is uncountable.

We conclude this section with another of Asplund's theorems.

Theorem 2.45 ([Asp]). *If the Banach space E is a subspace of a WCG space, then E is a weak Asplund space.*

Proof. By a deep renorming theorem due to Amir and Lindenstrauss (see, for instance, [Di] or [D-G-Z$_2$]) any WCG space admits an equivalent norm having strictly convex dual. (This obviously generalizes Exercise 2.40 (b).) Moreover, the new norm inherited by any subspace is easily seen also to have strictly convex dual, so Cor. 2.39 applies.

Remarks.

Theorem 2.11 was suggested by a remark in the Preiss-Zajicek paper [Pr-Z]; they proved a sharpened form of generic differentiability for continuous convex functions (in a Banach space with separable dual) and pointed out how it could be used to obtain a generic continuity theorem for monotone mappings. (We have replaced their term "α-angle porous" by "α-cone meager", feeling that the latter is a bit more descriptive.) They also pointed out that D. Gregory's argument could be used to give the proof of Theorem 2.14 (that Asplund spaces are "separably determined").

There is a great deal of material about monotone mappings and their applications in the two volumes by E. Zeidler [Ze] and in the book by D. Pascali and S. Sburlan [Pa-Sb], mostly for reflexive spaces. (The reader is warned that in the early sections the latter authors sometimes assume–without saying so–that their Banach spaces are reflexive.)

The fundamental Theorem 2.28 (on local boundedness of monotone operators) was first proved in slightly different form by Rockafellar [Ro$_2$]. The original proof has been considerably simplified by a number of authors; there is one due to P. M. Fitzpatrick (see [Pa-Sb] who have attributed to the latter author one of S. P. Fitzpatrick's early papers), and an even shorter proof, using the uniform boundedness theorem, in [Dr-L]. The one we present is a specialization to Banach spaces of a general result by J. Borwein and S. P. Fitzpatrick [Bor-F]. Recently, L. Veselý [private communication] has shown that *if $\overline{D(T)}$ is convex, then a maximal monotone T fails to be locally bounded at each point of* bdry $D(T)$. He first notes a Fact: If T is locally bounded at $x \in \overline{D(T)}$ (definition obvious) then, using maximal monotonicity and weak* compactness, $x \in D(T)$. Assume that $\overline{D(T)}$ is convex. If $x \in$ bdry $D(T) \subset \overline{D(T)}$ and T is locally bounded at x, then by the Fact, $x \in D(T)$. Suppose that x were in the boundary of $\overline{D(T)}$ and that there were a neighborhood U of x with $T(U)$ bounded. By the Bishop–Phelps theorem there would exist a point $z \in U \cap \overline{D(T)}$ and a nonzero element $w^* \in E^*$ such that $\langle w^*, z \rangle = \sup \langle w^*, D(T) \rangle$. Now, T would also be locally bounded at z, so (the Fact again) $z \in D(T)$ and we could choose $z^* \in T(z)$. The same kind of argument used just before Exercises 2.29 would show that $T(z)$ is unbounded, a contradiction. Since $x \notin$ bdry $\overline{D(T)}$, it must be in int $\overline{D(T)}$. By local boundedness, we can choose an open set U such that $x \in U \subset$ int $\overline{D(T)}$ and $T(U)$ is bounded. Thus, T is locally bounded at *every* point of U, which, by the Fact, implies that $U \subset D(T)$ and therefore $x \in$ int $D(T)$, another contradiction.

Section 3

3. Lower semicontinuous convex functions.

Our differentiability results for convex functions made heavy and consistent use of continuity, but in both the theoretical and applied aspects of convex functions, it is sometimes desirable to weaken this hypothesis. Lower semicontinuity is precisely what is needed. Uncomfortable as it may seem at first, the subject is best treated by introducing the seeming complication of admitting *extended* real-valued functions, that is, functions with values in $\mathbf{R} \cup \{\infty\}$. We adopt the conventions that

$$r \cdot \infty = \infty \text{ and } (-r) \cdot \infty = -\infty \text{ if } r > 0, \text{ and } r \pm \infty = \pm\infty \text{ for all } r \in \mathbf{R}.$$

(We won't have occasion to worry about $\infty - \infty$ or $0 \cdot \infty$.)

Definition 3.1 Let X be a Hausdorff space and let $f: X \rightarrow \mathbf{R} \cup \{\infty\}$. The *effective domain* of f is the set $\mathrm{dom}(f) = \{x \in X: f(x) < \infty\}$. We say that f is *lower semicontinuous* provided $\{x \in X: f(x) \leq r\}$ is closed in X for every $r \in R$. This is equivalent to saying that the *epigraph* of f

$$\mathrm{epi}(f) = \{(x, r) \in E \times \mathbf{R}: r \geq f(x)\}$$

is closed in $X \times \mathbf{R}$. Equivalently, f is lower semicontinuous provided

$$f(x) \leq \lim \inf f(x_\alpha)$$

whenever $x \in X$ and (x_α) is a net in X converging to x. We say that f is *proper* if $\mathrm{dom}(f) \neq \emptyset$. Our original definition of convexity for a real-valued function on a Banach space E applies without change.

Note that if f is convex, then so is $\mathrm{dom}(f)$. Also, *a function f is convex if and only if* $\mathrm{epi}(f)$ *is convex*. This last fact is important; it means that certain properties of lower semicontinuous convex functions can be deduced from properties of these (rather special) closed convex subsets of $E \times \mathbf{R}$. One can view this as saying that the study of lower semicontinuous convex functions is a special case of the study of closed convex sets.

3.2. Examples.

(a) Let C be a nonempty convex subset of E; then the *indicator function* δ_C, defined by $\delta_C(x) = 0$ if $x \in C$, $= \infty$ otherwise, is a proper convex function which is lower semicontinuous if and only if C is closed.

[This example is one reason for introducing extended real-valued functions, since it makes it possible to deduce certain properties of a closed convex set from properties of its lower semicontinuous convex indicator function. Thus, one can use this example to assert that the study of closed convex sets is a special case of the study of lower semicontinuous convex functions. It's all a matter of which viewpoint is more convenient, the geometrical or the analytical. It is best to be able to switch readily from one to the other.]

(b) Let A be any nonempty subset of E^* such that the weak* closed convex hull of A is not all of E^* (or, more simply, let A be a weak* closed convex proper subset of E^*) and let

$$\sigma_A(x) = \sup\{\langle x^*, x \rangle : x^* \in A\}, \quad x \in E;$$

then σ_A is a proper lower semicontinuous convex function, called the *support function* of A. (In Sec. 2 we assumed that A was bounded.)

(c) If f is a continuous convex function defined on a nonempty closed convex set C, extend f to be ∞ at the points of $E \setminus C$; the resulting function is a proper lower semicontinuous convex function.

The next proposition uses completeness of E to show where lower semicontinuous convex functions are necessarily continuous.

Proposition 3.3. *Suppose that f is a proper lower semicontinuous convex function on a Banach space E and that $D = \text{int dom}(f)$ is nonempty; then f is continuous on D.*

Proof. We need only show that f is locally bounded in D, since (as observed after Prop. 1.6) this implies that it is locally Lipschitzian in D. First, note that if f is bounded above (by M, say) in $B(x; \delta) \subset D$ for some $\delta > 0$, then it is bounded below in $B(x; \delta)$. Indeed, if y is in $B(x; \delta)$, then so is $2x - y$ and

$$f(x) \leq \frac{1}{2}[f(y) + f(2x - y)] \leq \frac{1}{2}[f(y) + M]$$

so $f(y) \geq 2f(x) - M$ for all $y \in B(x; \delta)$. Thus, to show that f is locally bounded in D, it suffices to show that it is locally bounded *above* in D. For each $n \geq 1$, let $D_n = \{x \in D : f(x) \leq n\}$. The sets D_n are closed and $D = \cup D_n$; since D is a Baire space, for some n we must have $U \equiv \text{int } D_n$ nonempty. We know that f is bounded above by n in U; without loss of generality, we can assume that $B(0; \delta) \subset U$ for some $\delta > 0$. If y is in D, with $y \neq 0$, then there exists $\mu > 1$ such that $z = \mu y \in D$ and hence (letting $0 < \lambda = \mu^{-1} < 1$), the set

$$V = \lambda z + (1 - \lambda)B(0; \delta) = y + (1 - \lambda)B(0; \delta)$$

is a neighborhood of y in D; Fig. 3.1 below illustrates the situation.

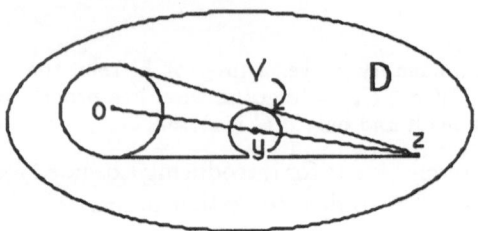

Fig. 3.1.

For any point $v = (1 - \lambda)x + \lambda z \in V$ (where $x \in B(0; \delta)$) we have

$$f(v) \leq (1 - \lambda)n + \lambda f(z),$$

so f is bounded above in V and the proof is complete.

3.4. Examples.

(a) The function f defined by $f(x) = 1/x$ on $(0, \infty)$, $f(x) = \infty$ on $(-\infty, 0]$ shows that f can be continuous at a boundary point x of dom(f) where $f(x) = \infty$. (Recall that the neighborhoods of ∞ in $(-\infty, \infty]$ are all the sets $(a, \infty]$, $a \in \mathbf{R}$.)

(b) Suppose that C is nonempty closed and convex; then the lower semicontinuous convex indicator function δ_C is continuous at $x \in C$ if and only if $x \in$ int C. Thus, *if* int $C = \emptyset$, *then* δ_C *is not continuous at any point of* $C = $ dom(δ_C).

Definition 3.5. Recall that if E is a Banach space, then so is $E \times \mathbf{R}$, under any norm which restricts to the original norm on the subspace E, for instance, $\|(x, r)\| = \|x\| + |r|$. Recall, also, that $(E \times \mathbf{R})^*$ can be identified with $E^* \times \mathbf{R}$, using the pairing

$$\langle (x^*, r^*), (x, r) \rangle = \langle x^*, x \rangle + r^* \cdot r.$$

Remark. If a proper lower semicontinuous convex function f is continuous at some point $x_0 \in$ dom(f), then dom(f) has nonempty interior and epi(f) has nonempty interior in $E \times \mathbf{R}$. (Indeed, $f(x) = \infty$ outside of dom(f), so x_0 cannot be a boundary point of the latter. Moreover, there exists an open neighborhood U of x_0 in dom(f) in which $f(x) < f(x_0) + 1$, so the open product set $U \times \{r : r > f(x_0) + 1\}$ is contained in epi(f).)

Definition 3.7. (a) The definition of the *subdifferential* ∂f for a proper lower semicontinuous convex function f is almost the same as that for a continuous function: If $x \in$ dom(f), define

$$\partial f(x) = \{x^* \in E^* : \langle x^*, y - x \rangle \leq f(y) - f(x) \text{ for all } y \in E\}$$

$$= \{x^* \in E^* : \langle x^*, y \rangle \leq f(x + y) - f(x) \text{ for all } y \in E\},$$

while $\partial f(x) = \emptyset$ if $x \in E \setminus$ dom(f). *It may also be empty at points of* dom(f), as shown in the first example below.

(b) If $x \in$ dom(f) we define $d^+ f(x)$ as before:

$$d^+ f(x)(y) = \lim_{t \to 0+} t^{-1}[f(x + ty) - f(x)], \qquad y \in E,$$

recognizing that $d^+ f(x)(y) = \infty$ if $x + ty \in E \setminus \text{dom}(f)$ for all $t > 0$. (It is also possible to have $d^+ f(x)(y) = -\infty$; consider, for instance, $d^+ f(0)(1)$ when $f(x) = -x^{1/2}$ for $x \geq 0, = \infty$ elsewhere.) The following important relationship is easily seen still to be valid in this more general situation: For any point $x \in \text{dom}(f)$,

$$x^* \in \partial f(x) \text{ if and only if } \langle x^*, y \rangle \leq d^+ f(x)(y) \text{ for all } y \in E.$$

It follows from this that for the example given above ($f(x) = -\sqrt{x}$ for $x \geq 0$), it must be true that $\partial f(0) = \emptyset$. In the first example below, one sees that it is possible to have $\partial f(x) = \emptyset$ for a dense set of points $x \in \text{dom}(f)$.

3.8 Examples.

(a) Let C be the closed (in fact, compact) convex subset of ℓ^2 defined by $C = \{x \in \ell^2 : |x_n| \leq 2^{-n}, n = 1, 2, 3, \ldots\}$ and define f on C by

$$f(x) = \Sigma[-(2^{-n} + x_n)^{1/2}].$$

Since each of the functions $x \to -(2^{-n} + x_n)^{1/2}$ is continuous, convex and bounded in absolute value by $2^{(-n+1)/2}$, the series converges uniformly, so f is continuous and convex. We claim that $\partial f(x) = \emptyset$ for any $x \in C$ such that $x_n > -2^{-n}$ for infinitely many n. Indeed, let e_n denote the n-th unit vector in ℓ^2. If $x^* \in \partial f(x)$ (so that, as noted above, $x^* \leq d^+ f(x)$), then for all n such that $x_n > -2^{-n}$, we have

$$-\|x^*\| \leq \langle x^*, e_n \rangle \leq d^+ f(x)(e_n) = -(1/2)(2^{-n} + x_n)^{-1/2},$$

an impossibility which implies that $\partial f(x) = \emptyset$. Note that if we make the usual extension (setting $f(x) = \infty$ for $x \in \ell^2 \setminus C$), then f is lower semicontinuous, but not continuous at any point of C ($= \text{bdry } C$).

(b) Let C be a nonempty closed convex subset of E; then for any $x \in C$, the subdifferential $\partial \delta_C(x)$ of the indicator function δ_C is the cone with vertex 0 of all $x^* \in E^*$ which "support" C at x, that is, which satisfy

$$\langle x^*, x \rangle = \sup\{\langle x^*, y \rangle : y \in C\} \equiv \sigma_C(x^*).$$

(Indeed, $x^* \in \partial \delta_C(x)$ if and only if $\langle x^*, y - x \rangle \leq \delta_C(y) - \delta_C(x)$, while x^* attains its supremum on C at x if and only if the left hand side of this latter inequality is at most 0, while the right side is always greater or equal to 0.)

(c) The following is an example of a function which is continuous and convex on ℓ_2 but is not bounded on the unit ball. Define $\phi(t)$ for $t \geq 0$ by $\phi(t) = 0$ if $0 \leq t \leq 1/2$ while $\phi(t) = 2t - 1$ if $t \geq 1/2$. For $x = (x_k) \in \ell_2$, define

$$f(x) = \sum_{n=1}^{\infty} \phi(\Sigma_{k \geq n} x_k^2).$$

(As the supremum of a sequence of continuous convex functions, f is lower semicontinuous and convex. Since the infinite sum actually has only finitely many nonzero terms, f is finite on all of ℓ_2, hence is continuous by Prop. 3.3. It tends to infinity on the sequence of norm-one elements of the form $x_1 = \frac{1}{\sqrt{3}} = x_n = x_{n+1}$ while $x_k = 0$ if $k \neq 1, n, n+1$.)

3.9. Exercises.

(a) A proper lower semicontinuous convex function f has a global minimum on E at x if and only if $0 \in \partial f(x)$.

(b) Suppose C is a nonempty convex subset of E and that f is a proper convex function on E with $\mathrm{dom}(f) \cap C \neq \emptyset$; then $f|_C$ has a minimum at the point $x \in C$ if and only if $0 \in \partial(f + \delta_C)(x)$.

Suppose that f is proper convex function on E. Define its dual function (or "Fenchel transform") f^* on E^* by

$$f^*(x^*) = \sup\{\langle x^*, x\rangle - f(x): x \in E\}, \qquad x^* \in E^*.$$

(c) Show that f^* is a proper lower semicontinuous extended real-valued convex function on E^*.

(d) Show that for all $x \in E$ and $x^* \in E^*$,

$$\langle x^*, x\rangle \leq f^*(x^*) + f(x),$$

with equality holding if and only if $x^* \in \partial f(x)$.

(e) Define f^{**} on E^{**} in the obvious way and show that its restriction to $E \subset E^{**}$ coincides with f if and only if the latter is lower semicontinuous.

(f) Compute f^* (that is, describe it as a function on E^* without explicit reference to f) for each of the following functions f:
  (i) $f_1(x) = \|x\|$.
  (ii) $f_2(x) = \frac{1}{2}\|x\|^2$.
  (iii) $f_3 = \delta_C$, where C is a nonempty closed convex subset of E.
  (iv) $f_4 = \sigma_A$, where A is a nonempty bounded convex subset of E^*.

Suppose that f and g are proper lower semicontinuous convex functions on E. Define their *inf-convolution* (or *epi-sum*) by

$$(f \nabla g)(x) = \inf\{f(y) + g(x - y): y \in E\}, \qquad x \in E.$$

(g) Show that $f \nabla g$ is a proper lower semicontinuous convex function on E with domain $\mathrm{dom}(f) + \mathrm{dom}(g)$.

(h) Given nonempty closed convex sets C_1 and C_2, compute $\delta_{C_1} \nabla \delta_{C_2}$.

(i) Show that if C is a nonempty convex set, then

$$(\delta_C \nabla \|\cdot\|)(x) = \mathrm{dist}(x, C).$$

(j) Using the definition of dual functions (above), show that $(f \nabla g)^* = f^* + g^*$.

(k) Show that if the convex set $\text{dom}(f) - \text{dom}(g)$ contains a neighborhood of the origin (in particular, if $\text{dom}(f) \cap \text{int dom}(g) \neq \emptyset$) then $(f + g)^* = f^* \nabla g^*$.

(f) Show that $f \nabla g$ is continuous at $x \in E$ if either f or g is continuous at x.

Definition 3.10. A point x of a subset X of E is said to be a *support point* of X provided there exists $x^* \in E^*$, $x^* \neq 0$, such that x^* attains its supremum on X at x. Any such x^* is said to *support* X *at* x, or be a *supporting functional* of X. (This is not to be confused with the support function σ_X of X, but there is an obvious relationship: x^* supports X at x if and only if $\sigma_X(x^*) = \langle x^*, x \rangle$.) The geometric terminology arises from the fact that a closed hyperplane is said to support X if one of its two closed half spaces contains X and if the hyperplane itself actually intersects X. If x^* supports X at x, then $H = \{y \in E : \langle x^*, y \rangle = \sigma_X(x^*)\}$ contains x and is just such a hyperplane.

It is an easy consequence of the Hahn-Banach theorem (or separation theorem) that *if a closed convex set C has nonempty interior, then every boundary point of C is a support point of C*. It is not obvious that a nonempty closed convex set with empty interior has any support points. Even if it does, there is a question as to how many support *functionals* it admits. For instance, even though every boundary point of the unit ball B of a Banach space E is a support point of B, there may be functionals on E which do not attain their suprema (that is, their norms) on B. This obviously cannot happen if E is reflexive (since B is then weakly compact) and, in fact, a deep theorem of R. C. James says that *if E is nonreflexive, then there exists an element of E^* which does not attain its supremum on B*. (See Diestel's Lecture Notes [Di] for a proof.) The fact that both the support points and support functionals of C are necessarily dense (in appropriate spaces) are the Bishop-Phelps theorems. Now, it is important to know that the subdifferential of a lower semicontinuous proper convex function f will be nonempty for at least *some* points of $\text{dom}(f)$. By applying the Bishop-Phelps techniques in $E \times \mathbf{R}$ to the closed convex epigraphs of lower semicontinuous convex functions, A. Brøndsted and R. T. Rockafellar showed that $\text{dom}(\partial f)$ is, in fact, dense in $\text{dom}(f)$. There are by now a number of approaches to the foregoing results and to Rockafellar's theorem that the subdifferential map is maximal monotone; our route will pass through key results by I. Ekeland and S. Simons.

Definition 3.11. If $0 < \lambda < 1$ define

$$K_\lambda = \{(x, r) \in E \times \mathbf{R} : \lambda \|x\| \leq -r\};$$

this is easily seen to be a closed convex cone (similar to the cone used in the Preiss-Zajíček theorem) which opens downward; it is the reflection through the origin of the epigraph of the function $\lambda \| \cdots \|$. Since the latter is continuous, K_λ has nonempty interior (containing $(0, -1)$, for instance).

The following lemma is an $E \times \mathbf{R}$ version of a classical maximality result due to Bishop and Phelps. It simply asserts that, in the partial ordering defined by K_λ, if a closed set A satisfies a certain boundedness condition, then any point of A is dominated by a *maximal* point of A.

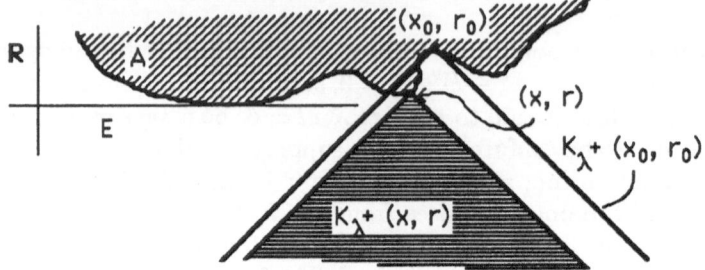

Fig. 3.2. This illustrates (3.1) and (3.2) below

Lemma 3.12. *Suppose that A is a closed nonempty subset of $E \times \mathbf{R}$, that $0 < \lambda < 1$ and that $\inf\{r : (x,r) \in A\} = 0$. For any point (x_0, r_0) in A, there exists $(x,r) \in A$ such that*

$$(x,r) \in A \cap [K_\lambda + (x_0, r_0)] \text{ and} \tag{3.1}$$

$$\{(x,r)\} = A \cap [K_\lambda + (x,r)]. \tag{3.2}$$

(Figure 3.2 above illustrates properties (3.1) and (3.2).)

Proof. It is convenient to define the continuous linear functional R on $E \times \mathbf{R}$ by $R(x,r) = r$. Choose $(x_1, r_1) \in A_0 \equiv A \cap [K_\lambda + (x_0, r_0)]$ such that

$$r_1 < \inf R(A_0) + 1.$$

Continuing by induction, we can choose a sequence $\{(x_n, r_n)\}$ such that

$$(x_{n+1}, r_{n+1}) \in A_n \equiv A \cap [K_\lambda + (x_n, r_n)] \text{ and}$$

$$r_{n+1} < \inf R(A_n) + 1/(n+1).$$

We claim that for every n,

(a) $A_{n+1} \subset A_n$ and   (b) diam $A_n \to 0$.

To see (a), note that

$$K_\lambda + (x_{n+1}, r_{n+1}) \subset K_\lambda + [K_\lambda + (x_n, r_n)] = K_\lambda + (x_n, r_n);$$

now interesect both sides with A to get the desired inclusion. To prove (b) it suffices to show that if $(y,s) \in A_n$, then $\|y - x_n\| + |s - r_n| \leq 1/n\lambda + 1/n$. Note that $s \geq \inf R(A_n) \geq \inf R(A_{n-1}) > r_n - 1/n$ (the second inequality following from (a)) and $\lambda\|y - x_n\| \leq r_n - s < 1/n$, so $\|y - x_n\| < 1/n\lambda$ and $0 \leq r_n - s < 1/n$. Since A and K_λ are closed, completeness guarantees that $\cap A_n$ is a single point, which we denote by (x,r). Since $(x,r) \in A_0$, we immediately have (3.1). To obtain (3.2), note that for each n,

$$K_\lambda + (x,r) \subset K_\lambda + [K_\lambda + (x_n, r_n)] = K_\lambda + (x_n, r_n),$$

so if $(y, s) \in A \cap [K_\lambda + (x, r)]$, then $(y, s) \in A_n$ for every n, hence we must have $(y, s) = (x, r)$.

[The proof given above goes through if E is merely a complete metric space; one simply replaces the partial ordering defined by K_λ by defining

$$(x, r) \leq (y, s) \text{ if and only if } \lambda d(x, y) \leq s - r.$$

The Banach space version is easier to visualize geometrically, and we have no need for the more general form.]

As an immediate consequence of this result we obtain Ekeland's *variational principle*, which can be viewed as saying that if $f(x_0)$ is nearly a minimum value for the lower semicontinuous function f, then a small Lipschitz continuous perturbation of f attains a strict minimum at a point z relatively close to x_0. (That is, there exists a Lipschitz continuous function g, with small Lipschitz constant, such that $f + g$ has a strict minimum at z.) This fact has found application in a wide variety of topics in nonlinear analysis; see Ekeland's survey in [Ek].

Lemma 3.13 (Ekeland). *Assume that f is a proper lower semicontinuous extended real-valued function on the Banach space E which is bounded below. Suppose that $\epsilon > 0$ and that $f(x_0) \leq \inf\{f(x): x \in E\} + \epsilon$. Then for any $\lambda > 0$ there exists a point $z \in \mathrm{dom}(f)$ such that*

(i) $\lambda\|z - x_0\| \leq f(x_0) - f(z),$ (ii) $\|z - x_0\| \leq \epsilon/\lambda$

and

(iii) $\lambda\|x - z\| + f(x) > f(z)$ whenever $x \neq z$.

Remark. Despite the fact that λ appears in the denominator, the estimate in (ii) need not be large for small λ; one can employ the great trick – to be used to good effect later – of taking $\lambda = \sqrt{\epsilon}$.

Proof. We assume without loss of generality that $\inf_E f = 0$, so we have $f(x_0) \leq \epsilon$. Put the equivalent norm $2\lambda\|\cdot\|$ on E and apply Lemma 3.12 to the closed set $A = \mathrm{epi}(f)$ and the cone $K_{1/2}$ (which we denote simply by K) to obtain a point (z, r) in $E \times \mathbf{R}$ such that

(1) $(z, r) \in A \cap [K + (x_0, f(x_0))]$ and (2) $\{(z, r)\} = A \cap [K + (z, r)]$.

From (1) we have $0 \leq f(z) \leq r < \infty$ and

$$\lambda\|z - x_0\| \leq f(x_0) - r \leq f(x_0) - f(z) \leq f(x_0) \leq \epsilon,$$

which yields assertions (i) and (ii). Assertion (iii) is obvious if $f(x) = \infty$. To see its validity in general, note first that if $f(z) < r$, then we must have $(z, r) \neq (z, f(z))$, so from (2) it follows that $(z, f(z))$ is not in $K + (z, r)$, that is, $0 > r - f(z)$, a contradiction. Thus, $r = f(z)$. From (2), again, if $f(x) < \infty$ and $x \neq z$, then $(x, f(x))$ is not in $K + (z, r)$, that is,

$$\lambda \|x - z\| > r - f(x) = f(z) - f(x),$$

which was to be shown.

The Brøndsted-Rockafellar theorem is an easy consequence of this lemma, but we first require a definition.

Definition 3.14. Let f be a proper convex lower semicontinuous function and suppose $x \in \text{dom}(f)$. For any $\epsilon > 0$ define the *ϵ-subdifferential* $\partial_\epsilon f(x)$ by

$$\partial_\epsilon f(x) = \{x^* : \langle x^*, y \rangle \leq f(x + y) - f(x) + \epsilon \text{ for all } y \in E\}.$$

It is clear that if $0 < \epsilon_1 < \epsilon_2$, then $\partial_{\epsilon_1} f(x) \subset \partial_{\epsilon_2} f(x)$. We show next that this set (necessarily convex and weak* closed) is always nonempty.

Proposition 3.15. *If f is a proper lower semicontinuous convex function on E, then $\partial_\epsilon f(x_0)$ is nonempty, for every $x_0 \in \text{dom}(f)$ and every $\epsilon > 0$.*

Proof. Figure 3.3 below shows immediately what is happening, but we must check some details.

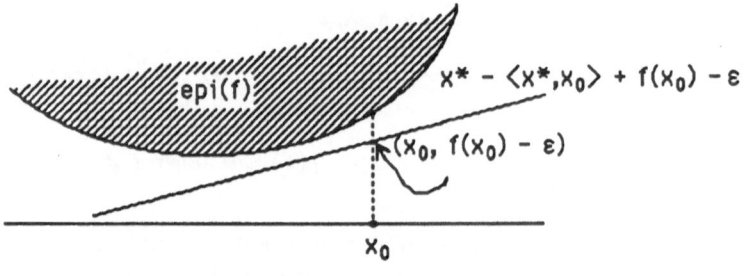

Fig. 3.3

Since $\text{epi}(f)$ is closed and convex and does not contain $(x_0, f(x_0) - \epsilon)$, there exists a linear functional $(y^*, r^*) \in E^* \times \mathbf{R}$ such that

$$\langle (y^*, r^*), (x_0, f(x_0) - \epsilon) \rangle < \langle (y^*, r^*), (x, r) \rangle, \quad x \in \text{dom}(f), \quad r \geq f(x),$$

that is

$$\langle y^*, x_0 \rangle + r^*[f(x_0) - \epsilon] < \langle y^*, x \rangle + r^* \cdot r \quad \text{if } x \in \text{dom}(f), \quad r \geq f(x).$$

Taking $x = x_0$ and $r = f(x_0)$ shows that $r^* > 0$. We can assume, in fact, that $r^* = 1$; simply replace y^* by y^*/r^*. If we then let $x^* = -y^*$ and take $r = f(x)$, we obtain the desired inequality.

We will also need the following fact, which is of substantial interest in its own right (especially in convex optimization).

Theorem 3.16. *Suppose that f and g are convex proper lower semicontinuous functions on the Banach space E and that there is a point in* $\mathrm{dom}(f) \cap \mathrm{dom}(g)$ *where one of them, say f, is continuous. Then*

$$\partial(f+g)(x) = \partial f(x) + \partial g(x), \qquad x \in \mathrm{dom}(f+g).$$

(The right side is the usual vector sum of sets.)

Remark. It is immediate from the definitions that for $x \in \mathrm{dom}(f+g)$ (which is identical to $\mathrm{dom}(f) \cap \mathrm{dom}(g)$), one must have

$$\partial f(x) + \partial g(x) \subset \partial(f+g)(x).$$

This inclusion can be proper. To see this, let $E = \mathbf{R}^2$, let f denote the indicator function δ_C and $g = \delta_L$, where C is the epigraph of the quadratic function $y = x^2$ and L is the x-axis. Obviously, C and L intersect only at the origin 0 and it is easily verified that $\partial f(0) = \mathbf{R}^- e$, where $\mathbf{e}$ is the vector $(0,1)$, and $\partial g(0) = \mathbf{R}e$, while

$$\partial(f+g)(0) = \mathbf{R}^2 \neq \partial f(0) + \partial g(0).$$

Proof. Suppose that $x_0^* \in \partial(f+g)(x_0)$. In order to simplify the argument, we can replace f and g by the functions

$$f_1(x) = f(x+x_0) - f(x_0) - \langle x_0^*, x \rangle \text{ and } g_1(x) = g(x+x_0) - g(x_0), \quad x \in E;$$

it is readily verified from the definitions that if $x_0^* \in \partial(f+g)(x_0)$, then $0 \in \partial(f_1 + g_1)(0)$ and if $0 \in \partial f_1(0) + \partial g_1(0)$, then $x_0^* \in \partial f(x_0) + \partial g(x_0)$. Without loss of generality, then, we assume that $x_0 = 0$, $x_0^* = 0$, $f(0) = 0$ and $g(0) = 0$. We want to conclude that 0 is in the sum $\partial f(0) + \partial g(0)$, under the hypothesis that $0 \in \partial(f+g)(0)$. This last means that

$$(f+g)(x) \geq (f+g)(0) = 0 \text{ for all } x \in E. \tag{3.3}$$

We now apply the separation theorem in $E \times \mathbf{R}$ to the two closed convex sets $C_1 = \mathrm{epi}(f)$ and $C_2 = \{(x,r): r \leq -g(x)\}$; this is possible because f has a point of continuity in $\mathrm{dom}(f) \cap \mathrm{dom}(g)$ and hence – recall Remark 3.6 – C_1 has nonempty interior. Moreover, it follows from (3.3) that C_2 misses the interior of $C_1 = \{(x,r): r > f(x)\}$. Since $(0,0)$ is common to both sets, it is contained in any separating hyperplane. Thus, there exists a functional $(x^*, r^*) \in E^* \times \mathbf{R}$, $(0,0) \neq (x^*, r^*)$, such that

$$\langle x^*, x \rangle + r^* \cdot r \geq 0 \text{ if } r \geq f(x) \text{ and } \langle x^*, x \rangle + r^* \cdot r \leq 0 \text{ if } r \leq -g(x).$$

Since $1 > f(0) = 0$ we see immediately that $r^* \geq 0$. To see that $r^* \neq 0$, (that is, that the separating hyperplane is not "vertical"), we argue by contradiction: If $r^* = 0$, then we must have $x^* \neq 0$; also $\langle x^*, x \rangle \geq 0$ for all $x \in \mathrm{dom}(f)$ and $\langle x^*, x \rangle \leq 0$ for all $x \in \mathrm{dom}(g)$. This says that x^* separates these two sets. This is impossible; by the continuity hypothesis, their intersection contains an interior point of dom(f). Without loss of generality, then, we can assume that $r^* = 1$ and hence, for any $x \in E$,

$$\langle -x^*, x - 0 \rangle \leq f(x) - f(0) \text{ and } \langle x^*, x - 0 \rangle \leq g(x) - g(0),$$

that is, $0 = -x^* + x^* \in \partial f(0) + \partial g(0)$, which completes the proof.

Theorem 3.17 (Brøndsted-Rockafellar). *Suppose that f is a convex proper lower semicontinuous function on the Banach space E. Then given any point $x_0 \in \text{dom}(f)$, $\epsilon > 0$, $\lambda > 0$ and any $x_0^* \in \partial_\epsilon f(x_0)$, there exist $x \in \text{dom}(f)$ and $x^* \in E^*$ such that*

$$x^* \in \partial f(x), \quad \|x - x_0\| \leq \epsilon/\lambda \text{ and } \|x^* - x_0^*\| \leq \lambda.$$

In particular, the domain of ∂f is dense in $\text{dom}(f)$.

Proof. By hypothesis, $\langle x_0^*, x - x_0 \rangle \leq f(x) - f(x_0) + \epsilon$ for all $x \in E$, so if we define
$$g(x) = f(x) - \langle x_0^*, x \rangle, \quad x \in E,$$
we see that g is proper and lower semicontinuous, with $\text{dom}(g) = \text{dom}(f)$. Moreover, $g(x_0) \leq \inf_E g + \epsilon$, so by Lemma 3.13 there exists $z \in \text{dom}(f)$ such that $\lambda \|z - x_0\| \leq \epsilon$ and $\lambda \|x - z\| + g(x) \geq g(z)$ for all $x \in E$. Letting $h(x) = \lambda \|x - z\|$ ($x \in E$), this last inequality implies that $0 \in \partial(g + h)(z) = \partial g(z) + \partial h(z)$ (by Theorem 3.16, since h is continuous). Thus, there exists $z^* \in \partial g(z) = \partial f(z) - x_0^*$ such that $-z^* \in \partial h(z) = \{x^* \in E^* : \|x^*\| \leq \lambda\}$. Let $x^* = z^* + x_0^*$ and $x = z$; then $x^* \in \partial f(x)$, $\|x^* - x_0^*\| \leq \lambda$ and $\|x - x_0\| \leq \epsilon/\lambda$, as required.

Some important special cases of the Bishop-Phelps density theorems are easy corollaries.

Theorem 3.18 (Bishop-Phelps). *Suppose that C is a nonempty closed convex subset of a Banach space E. Then*
  (i)   *The support points of C are dense in the boundary* bdry C *of C.*
  (ii)   *The support functionals of C are dense in the cone of all those functionals which are bounded above on C.*

Proof. (i). Suppose that $x_0 \in$ bdry C and that $0 < \epsilon < 1$. Let $f = \delta_C$ be the indicator function of C. Choose $x_1 \in E \setminus C$ such that $\|x_0 - x_1\| < \epsilon$ and apply the separation theorem to obtain $x_0^* \in E^*$, $\|x_0^*\| = 1$, such that $\sigma_C(x_0^*) < \langle x_0^*, x_1 \rangle$. This implies that for all $x \in C$

$$\langle x_0^*, x \rangle < \langle x_0^*, x_1 \rangle = \langle x_0^*, x_1 - x_0 \rangle + \langle x_0^*, x_0 \rangle \leq \epsilon + \langle x_0^*, x_0 \rangle$$

so $\langle x_0^*, x - x_0 \rangle \leq \epsilon = f(x) - f(x_0) + \epsilon$, that is $x_0^* \in \partial_\epsilon f(x_0)$. By Theorem 3.17 (taking $\lambda = \sqrt{\epsilon}$) there exist $x \in C = \text{dom}(f)$ and $x^* \in \partial f(x)$ (which says that x^* attains its supremum on C at x) such that

$$\|x - x_0\| \leq \sqrt{\epsilon} \quad \text{and} \quad \|x^* - x_0^*\| \leq \sqrt{\epsilon} < 1.$$

The last inequality implies that $x^* \neq 0$, so it is a supporting functional of C at x.

(ii). Suppose that $x_0^* \in E^*$ is such that $x_0^* \neq 0$ and $\sigma_C(x_0^*) < \infty$. Given $0 < \epsilon < \|x_0^*\|^2$, choose $x_0 \in C$ such that $\langle x_0^*, x_0 \rangle > \sigma_C(x^*) - \epsilon$. If we again let $f = \delta_C$, then

$$\langle x_0^*, x - x_0 \rangle < \epsilon = f(x) - f(x_0) + \epsilon \text{ for all } x \in C,$$

that is, $x_0^* \in \partial_\epsilon f(x_0)$. By Theorem 3.17 (with $\lambda = \sqrt{\epsilon}$ again), there exist $x \in \text{dom}(f) = C$ and $x^* \in \partial f(x)$ such that $\|x^* - x_0^*\| \leq \sqrt{\epsilon} < \|x_0^*\|$. This last inequality implies that $x^* \neq 0$, so the proof is complete.

Note that assertion (ii) (above) can be reformulated as a variational principle: *If the continuous linear functional x_0^* is bounded above on C, then there exists a continuous linear functional y^* of small norm (namely, $y^* = x^* - x_0^*$) such that $x_0^* + y^*$ attains its maximum on C.*

The version of the Bishop-Phelps theorem most frequently applied in the theory of Banach spaces is the following special case of Theorem 3.20(ii) (above); it is what most authors mean when they refer to the "Bishop-Phelps theorem".

Theorem 3.19. *Let E be a Banach space. Then the set of all functionals x^* in E^* which attain their norms on the unit ball, that is, which satisfy*

$$\langle x^*, x \rangle = \|x^*\| \text{ for some } x \in E \quad \text{with} \quad \|x\| = 1,$$

is norm dense in E^*. Equivalently, the duality map J has dense range.

There are much more general versions of the Bishop-Phelps theorems in the original paper [Bi-Ph], but they have not found wide application. For instance, one can prove that if a functional strictly separates C from a nonempty bounded set X, then it can be approximated by a functional which supports C and strictly separates C from X. An interesting special case of this result (taking X to be a single point) has the following proposition as an immediate consequence; as we will see, it can also be proved using the present methods.

Proposition 3.20. *A nonempty closed convex subset C of a Banach space E is the intersection of all the closed half-spaces defined by its supporting hyperplanes.*

Proof. We must show that if $y \in E \setminus C$, then there is a support functional of C which separates C from y. Let $d = \text{dist}(y, C)$ and use the separation theorem to choose $x_0^* \in E^*$ of norm 1 which separates C from the ball $B(y; d)$; this implies that $\sigma_C(x_0^*) = \langle x_0^*, y \rangle - d$.

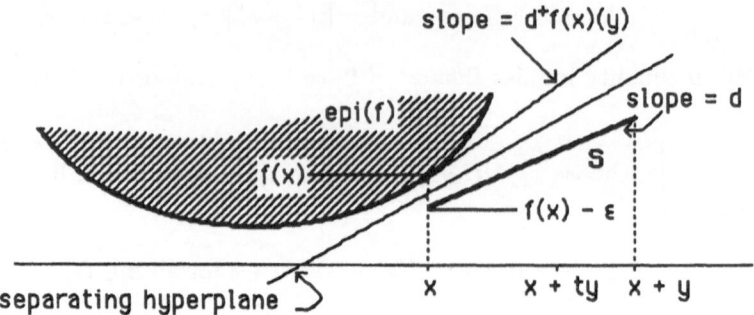

Fig. 3.4. The functional x_0* separates C and $B(y;d)$

Let f denote the indicator function δ_C and choose $\epsilon > 0$ sufficiently small such that $\sqrt{\epsilon}(\sqrt{\epsilon} + d + \epsilon) < d/2$. Next, choose $x_0 \in C \cap B(y; d + \epsilon)$, so $\langle x_0^*, y - x_0 \rangle \le \|y - x_0\| \le d + \epsilon$. For any $x \in C$ we have

$$\langle x_0^*, x - x_0 \rangle = \langle x_0^*, x - y \rangle + \langle x_0^*, y - x_0 \rangle \le$$

$$\le \sigma_C(x_0^*) - \langle x_0^*, y \rangle + d + \epsilon = \epsilon;$$

that is, $x_0^* \in \partial_\epsilon f(x_0)$. By Theorem 3.17 (taking $\lambda = \sqrt{\epsilon}$) there exist elements $x_\epsilon \in \mathrm{dom}(f) \equiv C$ and $x_\epsilon^* \in \partial f(x_\epsilon)$ (that is, x_ϵ^* supports C at x_ϵ) such that $\|x_\epsilon - x_0\| \le \sqrt{\epsilon}$ and $\|x_\epsilon^* - x_0^*\| \le \sqrt{\epsilon}$. It follows that for all $x \in C$,

$$\langle x_\epsilon^*, x - y \rangle \le \langle x_\epsilon^*, x_\epsilon - y \rangle = \langle x_\epsilon^* - x_0^*, x_\epsilon - y \rangle + \langle x_0^*, x_\epsilon - y \rangle \le$$

$$\le \|x_\epsilon^* - x_0^*\| \cdot (\|x_\epsilon - x_0\| + \|x_0 - y\|) + \sigma_C(x_0^*) - \langle x_0^*, y \rangle \le$$

$$\le \sqrt{\epsilon}(\sqrt{\epsilon} + d + \epsilon) - d < -d/2,$$

so $\sigma_C(x_\epsilon^*) < \langle x_\epsilon^*, y \rangle$.

Corollary 3.21. *Suppose that f is a convex lower semicontinuous proper function on E; then f is the upper envelope of the continuous affine functions defined by its subdifferentials, that is, for any x in $\mathrm{dom}(f)$,*

$$f(x) = \sup\{\langle y^*, x - y \rangle + f(y) : y^* \in \partial f(y) \text{ for some } y \in \mathrm{dom}(\partial f)\},$$

Proof. If $x \in \mathrm{dom}(f)$ and $\epsilon > 0$, apply Prop. 3.20 to obtain a closed hyperplane in $E \times \mathrm{R}$ which supports $C = \mathrm{epi}\,(f)$ (at $(y, f(y))$, say) and misses the point $(x, f(x) - \epsilon)$, as shown in Fig. 3.5 below.

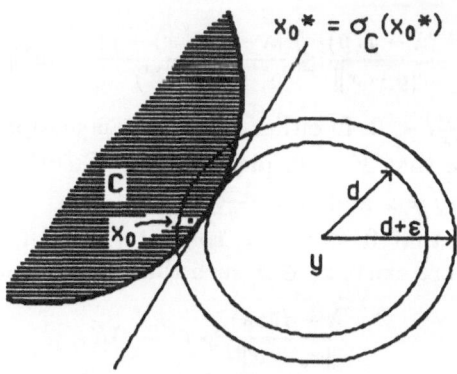

Fig. 3.5

By using the same reasoning as in the proof of Prop. 3.15, this defines an element $y^* \in E^*$ which is in $\partial f(y)$ and which satisfies

$$\langle y^*, x - y \rangle + f(y) > f(x) - \epsilon,$$

precisely what is needed.

The next two lemmas lead easily to S. Simons' proof of Rockafellar's maximal monotonicity theorem for subdifferentials.

Lemma 3.22. *Suppose that f is a lower semicontinuous proper convex function on E. If $\alpha, \beta > 0$, $x_0 \in E$ and $f(x_0) < \inf_E f + \alpha\beta$, then there exist $x \in E$ and $x^* \in \partial f(x)$ such that $\|x - x_0\| < \beta$ and $\|x^*\| < \alpha$.*

Proof. Choose $\epsilon > 0$ such that $f(x_0) - \inf_E f < \epsilon < \alpha\beta$ and then choose λ such that $\epsilon/\beta < \lambda < \alpha$. It follows that $0 \in \partial_\epsilon f(x_0)$ so by Theorem 3.17, there exist $x \in \text{dom}(f)$ and $x^* \in \partial f(x)$ such that $\|x^*\| \leq \lambda < \alpha$ and $\|x - x_0\| \leq \epsilon/\lambda < \beta$.

Lemma 3.23. *With f as in the previous lemma, suppose that $x \in E$ (not necessarily in $\text{dom}(f)$) and that $\inf_E f < f(x)$. Then there exist $z \in \text{dom}(f)$ and $z^* \in \partial f(z)$ such that*

$$f(z) < f(x) \quad \text{and} \quad \langle z^*, x - z \rangle > 0.$$

Proof. Fix $\lambda \in \mathbf{R}$ such that $\inf_E f < \lambda < f(x)$ and let

$$K = \sup_{y \in E, y \neq x} \frac{\lambda - f(y)}{\|y - x\|}.$$

We first show that $0 < K < \infty$. To that end, let $F = \{y \in E : f(y) \leq \lambda\}$, so F is closed, nonempty and $x \notin F$. Since $\text{dom}(f) \neq \emptyset$, one can apply the separation theorem in $E \times \mathbf{R}$ to find $u^* \in E^*$ and $r \in \mathbf{R}$ such that $f \geq u^* + r$. Suppose that $y \in E$ and that $y \neq x$. If $y \in F$, then

$$\lambda - f(y) \leq \lambda - \langle u^*, y \rangle - r \leq |\lambda - \langle u^*, x \rangle - r| + \langle u^*, x - y \rangle$$

hence
$$\frac{\lambda - f(y)}{\|y - x\|} \le \frac{|\lambda - \langle u^*, x \rangle - r|}{\mathrm{dist}(x, F)} + \|u^*\|.$$

If $y \notin F$, then $\frac{\lambda - f(y)}{\|y-x\|} < 0$. In either case, there is an upper bound for $\frac{\lambda - f(y)}{\|y-x\|}$, so $K < \infty$. To see that $K > 0$, pick any $y \in E$ such that $f(y) < \lambda$. Since $\lambda < f(x)$, we have $y \ne x$ and $K \ge \frac{\lambda - f(y)}{\|y-x\|} > 0$.

Suppose, now, that $0 < \epsilon < 1$, so that $(1 - \epsilon)K < K$ and hence, by definition of K, there exists $x_0 \in E$ such that $x_0 \ne x$ and
$$\frac{\lambda - f(x_0)}{\|x_0 - x\|} > (1 - \epsilon)K.$$

For $z \in E$, let $N(z) = K\|z - x\|$; we have shown that $(1-\epsilon)N(x_0)+f(x_0) < \lambda$, that is, $(N + f)(x_0) < \lambda + \epsilon N(y)$. We claim that $\lambda \le \inf_E(N + f)$ Indeed, if $z = x$, then we have $\lambda < f(x) = (N + f)(z)$, while if $z \ne x$, then $\frac{\lambda - f(z)}{\|z-x\|} \le K$, from which it follows that $\lambda \le (N + f)(z)$. Thus, we have shown that there is a point $x_0 \in E$, $x_0 \ne x$, such that
$$(N + f)(x_0) < \inf_E(N + f) + \epsilon K \|x_0 - x\|.$$

We now apply Lemma 3.22 to $N + f$, with $\beta = \|x_0 - x\|$ and $\alpha = \epsilon K$. Thus, there exists $z \in \mathrm{dom}(N + f) \equiv \mathrm{dom}(f)$ and $w^* \in \partial(N + f)(z)$ such that $\|z - x_0\| < \|x - x_0\|$ and $\|w^*\| < \epsilon K$. It follows that $\|z - x\| > 0$. From the sum formula (Theorem 3.16),
$$\partial(N + f)(z) = \partial N(z) + \partial f(z),$$

so there exist $y^* \in \partial N(z)$ and $z^* \in \partial f(z)$ such that $w^* = y^* + z^*$. Since $y^* \in \partial N(z)$, we must have $\langle y^*, z - x \rangle \ge N(z) - N(x) = K\|z - x\|$. Thus
$$\langle z^*, x - z \rangle = \langle y^*, z - x \rangle + \langle w^*, x - z \rangle \ge K\|z - x\| - \|w^*\| \cdot \|x - z\| >$$
$$> (1 - \epsilon)K\|z - x\| > 0.$$

Since $z^* \in \partial f(z)$, we have $f(x) \ge f(z) + \langle z^*, x - z \rangle > f(z)$, which completes the proof.

Theorem 3.24 (Rockafellar). *If f is a proper lower semicontinuous convex function on a Banach space E, the its subdifferential ∂f is a maximal monotone operator.*

Proof. Suppose that $x \in E$, that $x^* \in E^*$ and that $x^* \notin \partial f(x)$. Thus, $0 \notin \partial(f - x^*)(x)$, which implies that $\inf_E(f - x^*) < (f - x^*)(x)$. By Lemma 3.23 there exists $z \in \mathrm{dom}(f - x^*) \equiv \mathrm{dom}(f)$ and $z^* \in \partial(f - x^*)(z)$ such that $\langle z^*, z - x \rangle < 0$. Thus, there exists $y^* \in \partial f(z)$ such that $z^* = y^* - x^*$, so that $\langle y^* - x^*, z - x \rangle < 0$.

In Section 2 we noted that the subdifferential ∂f of a continuous convex function f is cyclically monotone, but we did not need the continuity hypothesis: If f is a convex proper lower semicontinuous function on E, if $x_0, x_1, \ldots, x_n = x_0$ are in $D(\partial f)$ and if $x_k^* \subset \partial f(x_k)$, $k = 1, 2, \ldots, n$, then

$$\Sigma_{k=1}^n \langle x_k^*, x_k - x_{k-1} \rangle \geq \Sigma [f(x_k) - f(x_{k-1})] = f(x_n) - f(x_0) = 0.$$

Thus, ∂f is both cyclically monotone and maximal monotone.

Definition 3.25. A monotone operator T is said to be *maximal cyclically monotone* provided $T = S$ whenever S is cyclically monotone and $G(T) \subset G(S)$. Clearly, a maximal monotone operator which is cyclically monotone is necessarily maximal cyclically monotone.

It is an interesting fact that *subdifferentials are the only maximal cyclically monotone operators.*

Proposition 3.26 (Rockafellar). *If $T: E \to 2^{E^*}$ is maximal cyclically monotone, with $D(T) \neq \emptyset$, then there exists a proper convex lower semicontinuous function f on E such that $T = \partial f$.*

Proof. Fix $x_0 \in D(T)$ and $x_0^* \in T(x_0)$. For $x \in E$ define

$$f(x) = \sup\{\langle x_n^*, x - x_n \rangle + \langle x_{n-1}^*, x_n - x_{n-1} \rangle + \cdots + \langle x_0^*, x_1 - x_0 \rangle\}$$

where the supremum is taken over all finite sets of elements $x_k \in D(T)$ and $x_k^* \in T(x_k)$, $k = 1, 2, \ldots, n$, $n = 1, 2, 3, \ldots$. Since f is the pointwise supremum of a family of continuous affine functions, it is convex and lower semi continuous and $f(x) > -\infty$ for all x. To see that f is proper, we can use monotonicity of T to show that $f(x_0) \leq 0$. Indeed, given any sum of terms of the form entering into the definition of $f(x_0)$, let $y_k = x_{n-k}$ and $y_k^* = x_{n-k}^*$, $k = 0, 1, \ldots, n$. The resulting sum is now the negative of a typical cyclic sum, hence is at most equal to 0. By the cyclic maximality of T, to conclude that $T = \partial f$ we need only show that $G(T) \subset G(\partial f)$. Suppose, then, that $x \in D(T)$ and $x^* \in T(x)$. We will have $(x, x^*) \in G(\partial f)$ if we can show that

$$\langle x^*, y - x \rangle \leq f(y) - \lambda \text{ for all } y \in E, \text{ whenever } \lambda < f(x).$$

(Note that by taking $y = x_0$, this will imply that $-\lambda \geq \langle x^*, x_0 - x \rangle$ whenever $\lambda < f(x)$, which shows that $f(x) < \infty$.) Now, by definition, there exist $x_k \in D(T)$ and $x_k^* \in T(x_k)$, $k = 1, 2, \ldots, n$, such that

$$\lambda < \langle x_n^*, x - x_n \rangle + \langle x_{n-1}^*, x_n - x_{n-1} \rangle + \cdots + \langle x_0^*, x_1 - x_0 \rangle.$$

Let $x_{n+1} = x$, $x_{n+1}^* = x^*$. For any $y \in E$

$$f(y) \geq \langle x_{n+1}^*, y - x_{n+1} \rangle + \langle x_n^*, x_{n+1} - x_n \rangle + \cdots + \langle x_0^*, x_1 - x_0 \rangle >$$

$$> \langle x_{n+1}^*, y - x_{n+1} \rangle + \lambda = \langle x^*, y - x \rangle + \lambda,$$

which completes the proof.

It is obvious that if f and g are proper lower semicontinuous functions which only differ by an additive constant, then $\partial f = \partial g$. Rockafellar [Ro₃] (see also [Tay]) has proved the converse; in particular, this shows that *if T is a maximal cyclically monotone operator* (so that $T = \partial f$ for an appropriate f), *then – within an additive constant – the latter is unique.*

Consider the following assertion concerning two maximal monotone operators S and T:

$$\text{If } D(S) \cap \operatorname{int} D(T) \neq \emptyset, \text{ then } S + T \text{ is maximal monotone.} \qquad (3.4)$$

(By definition, $D(S+T) = D(S) \cap D(T)$). Theorem 3.16 (that the subdifferential of the sum of two convex functions is the sum of their subdifferentials) shows that *(3.4) is true whenever $S = \partial f$ and $T = \partial g$, where f and g are proper lower semicontinuous convex functions.* Indeed, if $\operatorname{int} D(T) \neq \emptyset$, then [since $\operatorname{int} \operatorname{dom}(\partial g) \subset \operatorname{int} \operatorname{dom}(g)$ and g is continuous on $\operatorname{int} \operatorname{dom}(g)$] we conclude that g is continous at some point of $D(S) \cap D(T) \subset \operatorname{dom}(f) \cap \operatorname{dom}(g)$, so Theorem 3.16 implies that $S+T = \partial(f+g)$ and Theorem 3.24 implies that the latter is maximal monotone. The remark following Theorem 3.16 shows that, even in a two dimensional Banach space, the conclusion of (3.4) can fail if $D(S) \cap \operatorname{int} D(T) = \emptyset$; in that example, the graph of $\partial f + \partial g$ is strictly contained in that of the maximal monotone operator $\partial(f + g)$.

Rockafellar [Ro₅] has shown that (3.4) is indeed true for arbitrary maximal monotone operators S and T *provided E is assumed to be reflexive*, a fact that has been very useful in nonlinear analysis. Its validity for nonreflexive spaces remains an interesting open question.

We conclude this section by taking a brief look at a class of maximal monotone operators which do *not* arise from convex functions (that is, are not cyclic), but which do arise as subdifferentials of a certain class of functions.

Definition 3.27. (a) Let E and F be Banach spaces. We say that a mapping $K : E \times F \to \mathbf{R} \cup \{\pm\infty\}$ is a *saddle function* provided it is concave in the first variable and convex in the second; more precisely, we require that the function $x \to -K(x,y)$ [resp. $y \to K(x,y)$] be a convex function for each fixed $y \in E$ [resp. each fixed $x \in E$].

(b) The *domain* or *effective domain* $\operatorname{dom}(K)$ of K is defined to be the set of points (x,y) where K is finite-valued and

$$K(x',y) < \infty \text{ for all } x' \in E \text{ and } K(x,y') > -\infty \text{ for all } y' \in E.$$

We say that K is *proper* if its domain is nonempty.

(c) If K is a proper saddle function we define its *subdifferential* ∂K in the following way: Let $\partial_1 K(x,y) \subset E^*$ be the subdifferential of the *convex* function $-K(\cdot,y)$ at the point $(x,y) \in \operatorname{dom}(K)$ and let $\partial_2 K(x,y) \subset F^*$ be the subdifferential of $K(x,\cdot)$ at the same point. For any (x,y) in $E \times F$ define

$$\partial K(x,y) = \partial_1 K(x,y) \times \partial_2 K(x,y) \subset E^* \times F^*.$$

Note that if we put any reasonable norm on $E \times F$, then it is a Banach space (take, for instance, $\|(x,y)\| = \|x\| + \|y\|$) and its dual can be identified with $E^* \times F^*$, using the pairing

$$\langle (x^*, y^*), (x,y) \rangle = \langle x^*, x \rangle + \langle y^*, y \rangle.$$

3.28 Examples.

(a) Let $E = F$ be Hilbert space and define

$$K(x,y) = (1/2)[\|y\|^2 - \|x\|^2].$$

In view of the fact that the derivative of $(1/2)\| \cdot \|^2$ at u is the functional $\langle u, \cdot \rangle$, that is, can be identified with u itself, we see that

$$\partial K(x,y) = \{(x,y)\}.$$

(b) Again, let $E = F$ be Hilbert space and define

$$K(x,y) = \|y\|^2 - \langle x, y \rangle.$$

Here we have $\partial_1 K(x,y) = \{y\}$ and $\partial_2 K(x,y) = \{2y - x\}$, so

$$\partial K(x,y) = \{(y, 2y - x)\}.$$

Proposition 3.29. *If* $K: E \times F \to \mathbf{R} \cup \{\pm\infty\}$ *is a proper saddle function, then its subdifferential* $\partial K: E \times F \to 2^{(E \times F)^*}$ *is monotone.*

Proof. We must show that

$$\langle (x_1^*, y_1^*) - (x_2^*, y_2^*), (x_1, y_1) - (x_2, y_2) \rangle \geq 0$$

whenever $(x_1, y_1), (x_2, y_2) \in E \times F$ and that $(x_i^*, y_i^*) \in \partial K(x_i, y_i)$, $i = 1, 2$. Equivalently, given the definition of the pairing between $E \times F$ and $E^* \times F^*$, we want

$$\langle x_1^* - x_2^*, x_1 - x_2 \rangle + \langle y_1^* - y_2^*, y_2 - y_1 \rangle \geq 0,$$

that is,

$$\langle x_1^* - x_2^*, x_2 - x_1 \rangle + \langle y_2^* - y_1^*, y_1 - y_2 \rangle \leq 0. \tag{3.5}$$

Since $x_i^* \in \partial_1 K(x_i, y_i)$ and $y_i^* \in \partial_2 K(x_i, y_i)$, $i = 1, 2$, we have for all points $(x,y) \in E \times F$

$$\langle x_1^*, x - x_1 \rangle \leq -K(x, y_1) + K(x_1, y_1), \tag{3.6}$$

$$\langle x_2^*, x - x_2 \rangle \leq -K(x, y_2) + K(x_2, y_2), \tag{3.7}$$

$$\langle y_1^*, y - y_1 \rangle \leq K(x_1, y) - K(x_1, y_1), \tag{3.8}$$

$$\langle y_2^*, y - y_2 \rangle \leq K(x_2, y) - K(x_2, y_2). \tag{3.9}$$

We want to evaluate these four inequalities at x_2, x_1, y_2 and y_1 respectively and then add them to get (3.5) on the left side and 0 on the right side, but we must first check that the resulting cancellations on the right side do not involve adding ∞ to $-\infty$. By choosing (x, y) to be any element of dom(K), we know–by definition–that $-K(x, y_1) < \infty$, so (1) implies that $K(x_1, y_1) > -\infty$. Also, $K(x_1, y) < \infty$, so (3.8) implies that $K(x_1, y_1) < \infty$. Similar reasoning using (3.7) and (3.9) shows that $K(x_2, y_2)$ is also finite. Now, carry out the substitutions indicated at the beginning of this paragraph; it follows that $K(x_2, y_1)$ and $K(x_1, y_2)$ are also finite valued and we can now sum all four inequalities to get the desired result.

Under appropriate topological hypotheses on K (that it be proper and "closed" in a certain reasonable sense) ∂K will be *maximal* monotone; see [Ro$_4$].

Definition 3.30. Let E be a Banach space and suppose that K is a proper saddle function on $E \times E$. Define $T_K : E \to 2^{E^*}$ by

$$x^* \in T_K(x) \text{ provided } (x^*, x^*) \in \partial K(x, x).$$

The graph $G(T_K)$ of T_K is readily seen to be monotone, since it is just the intersection of $G(\partial K)$ with the linear subspace $D \times D^*$ of the product space $(E \times E) \times (E^* \times E^*)$, where D [resp. D^*] is the diagonal in $E \times E$ [resp. $E^* \times E^*$].

3.31 Example.

Each of the saddle functions K in Example 3.28 induces the same monotone operator T_K, since $T_K(x) = x$ for all x in each case.

In recent work, E. Krauss [Kr$_1$] (see also [Kr$_2$] and references therein) has shown that *every maximal monotone operator $T : E \to 2^{E^*}$ is of the form $T = T_K$ for some proper closed saddle function K on $E \times E$*. The previous example shows that K need not be unique, and Krauss has devoted considerable effort to showing how K can be chosen to have certain special properties. A different representation of general monotone operators on E (as subdifferentials of convex functions on $E \times E^*$) has been investigated by S. Fitzpatrick [Fi$_2$], who poses a number of related open problems.

Remarks.

The elementary properties of lower semicontinuous convex functions and their subdifferentials play a fundamental role in convex analysis (an approach to the calculus of variations and optimization which replaces differentiable functions by convex functions). Thus, any contemporary text on that subject will have some overlap with the early portions of this section (and with the first section). See, for instance, Aubin and Ekeland [Au-Ek] and Ekeland and Temam [Ek-T].

There was some reluctance on our part to abandon the original easily visualized geometric approach to the Bishop-Phelps theorems in favor of the more "analytic" approach, using Ekeland's variational principle. (M. Fabian [Fa$_2$] has reversed this approach, showing how to deduce Ekeland's principle and the Brøndsted-Rockafellar theorem (3.17) from a Bishop-Phelps lemma.) In the first edition of these notes [Ph$_4$] we stayed with the variational approach in part because it was a key step in a slightly cumbersome but very useful result by J. Borwein [Bor₁] which led to the Brøndsted-Rockafellar theorem and, especially, to a new proof of Rockafellar's theorem (3.24) on the maximal monotonicity of the subdifferential map (of a lower semicontinuous proper convex function). The recent, dramatically simpler proof of the latter theorem by S. Simons [Si] changed all that. While it still seems desirable first to prove Ekeland's variational principle (since it is the most immediate and fundamental example of a perturbed optimization theorem), it is now possible to omit a great deal of material which had been preparatory to Rockafellar's theorem. Most books which present the latter avoid giving a proof (referring the reader to Rockafellar's second of three proofs [Ro₃]) and at least one book reproduces Rockafellar's incorrect first proof.

Section 4

4. Smooth variational principles, Asplund spaces, weak Asplund spaces.

It is clear that Ekeland's variational principle (Lemma 3.13) is an extremely useful form of the "maximality points lemma" (3.12); it was a key step in a sequence of fundamental results. As shown in Ekeland's survey article [Ek], it has found application in such diverse areas as fixed-point theorems, non-linear semigroups, optimization, mathematical programming, control theory and global analysis. Recall the statement:

If f is lower semicontinuous on E, $\epsilon > 0$ and x_0 is such that $f(x_0) \leq \inf_E f + \epsilon$, then for any $\lambda > 0$ there exists $v \in E$ such that

$$\lambda \|x_0 - v\| \leq f(x_0) - f(v) \leq \epsilon \text{ and } f(x) + \lambda \|x - v\| > f(v) \text{ whenever } x \neq v.$$

One drawback to this result is that, even if f be differentiable, the perturbed version $f + \lambda \|(\cdot) - v\|$ will not be differentiable at v. This objection was first overcome by J. Borwein and D. Preiss [Bor-P] (see Theorem 4.20 below). A substantially simpler version was later obtained by R. Deville, G. Godefroy and V. Zizler [D-G-Z$_{1,3}$], and that is what we present next. In order to handle Gâteaux and Fréchet differentiability (and other kinds) simultaneously, we require the following definitions.

Definition 4.1.
   (a) A *bornology* on E, denoted by β, will be any family of bounded sets S whose union is all of E, which is closed under reflection through the origin (that is, $S \in \beta$ implies $-S \in \beta$), under multiplication by positive scalars and is directed upwards (that is, the union of any two members of β is contained in some member of β). There are many possibilities, but the following choices for β are of main interest to us: (i) Denote by $\beta = G$ the *Gâteaux* bornology consisting of all finite symmetric sets. (ii) By H we denote the *Hadamard* bornology, consisting of all compact symmetric sets. (iii) Let W denote the *weak Hadamard* bornology, consisting of all weakly compact symmetric sets. (iv) Finally, F denotes the *Fréchet* bornology consisting of all bounded symmetric sets. It is clear that G and F are the smallest and largest possible bornologies.

(b) A real-valued function f is said to be β-*differentiable* at x and $x^* \in E^*$ is called its β-*derivative* at x, if for each S in β,

$$\frac{f(x+ty) - f(x)}{t} - \langle x^*, y \rangle \to 0 \text{ as } t \to 0^+,$$

uniformly for $y \in S$. The symmetry and directed properties of a bornology imply that this right-hand limit is in fact a two-sided limit. We denote the β-derivative of f at x by $\nabla_\beta f(x)$; it is clear that, in terms of our earlier notation, $\nabla_G f(x) = df(x)$ and $\nabla_F f(x) = f'(x)$.

Definition 4.2. A *bump function* on E is a real-valued function ϕ which is bounded and has bounded nonempty support $supp(\phi) = \{x \in E : \phi(x) \neq 0\}$. We will say that the Banach space E *has property* (H_β) provided there exists on E a bump function b which is β-differentiable and globally Lipschitzian. It is straightforward to verify that a β-differentiable function f is Lipschitzian if and only if its β-derivative $x \to \nabla_\beta f(x)$ is bounded on E.

Proposition 4.3. *If E admits an equivalent β-differentiable norm (at nonzero points), then it necessarily has property (H_β).*

Proof. Let $\psi: \mathbf{R} \to \mathbf{R}$ be any C^1 function with bounded derivative and nonempty support contained in the interval $[\frac{1}{2}, \frac{3}{2}]$ and define $\phi(x) = \psi(\|x\|)$; this is β-differentiable and vanishes if $\|x\| < 1/2$ or $\|x\| > 3/2$. Since the derivative of the norm (at nonzero points) always has norm one, the chain rule shows that ϕ has bounded derivative.

The converse to this proposition fails dramatically: R. Haydon [Ha$_2$] has constructed a compact Hausdorff space X with the property that *there exists a Lipschitzian Fréchet differentiable bump function on $C(X)$ but the latter does not even admit an equivalent Gâteaux differentiable norm.*

Definition 4.4. A function $f: E \to (-\infty, \infty]$ attains a *strong minimum* at $y \in E$ if $f(y) = \inf_E f$ and $\|y_n - y\| \to 0$ whenever $y_n \in E$ and $f(y_n) \to f(y)$. If f is bounded on E, we define

$$\|f\|_\infty = \sup \{|f(x)| : x \in E\}.$$

The following general theorem of Deville, Godefroy and Zizler [DGZ$_3$] will yield some important corollaries, which are obtained by making judicious choices of the Banach space F.

Theorem 4.5. *Let E be a Banach space and F a Banach space of continuous bounded real-valued functions g on E such that*

(1) $\|g\|_\infty \leq \|g\|_F$ for all $g \in F$.

(2) For each $g \in F$ and $x \in E$, the function $y \to g_x(y) = g(x+y)$ is in F and $\|g_x\|_F = \|g\|_F$.

(3) For each $g \in F$ and $\alpha \in \mathbf{R}$, the function $y \to g(\alpha y)$ is in F.

(4) There exists a bump function in F.

If $f: E \to (-\infty, \infty]$ is proper, lower semicontinuous and bounded below, then the set G of all $g \in F$ such that $f + g$ attains a strong minimum on E is a dense G_δ subset of F.

Proof. It is helpful to introduce a notion which is analogous to the notion of a slice used in Section 2. If $g: E \to (-\infty, \infty]$ is lower semicontinuous and bounded below, we define, for any $\alpha > 0$, the closed set

$$S(g; \alpha) = \{x \in E: g(x) \le \inf_E g + \alpha\}.$$

It is easy to verify that if $\alpha > 0$ and g_1, g_2 are both bounded below and satisfy

$$g_1 \le g_2 + \alpha/3 \text{ and } g_2 \le g_1 + \alpha/3, \tag{4.1}$$

then $S(g_1; \alpha/3) \subset S(g_2; \alpha)$. Define

$$U_n = \{g \in F: \text{diam } S(f + g; \alpha) < 1/n, \text{ for some } \alpha > 0\}.$$

We will show that each of the sets U_n is dense and open in F and that their intersection is the desired set G. To see that U_n is open, suppose that $g \in U_n$, with a corresponding $\alpha > 0$. Then for any $h \in F$ such that $\|g - h\|_F < \alpha/3$, we have $\|g - h\|_\infty < \alpha/3$ and hence the functions $g_1 = f + g$ and $g_2 = f + h$ satisfy (4.1). Thus $S(f + h; \alpha/3) \subset S(f + g; \alpha)$ and therefore diam $S(f + h; \alpha/3) < 1/n$, so $h \in U_n$. To see that each U_n is dense in F, suppose that $g \in F$ and $\epsilon > 0$; it suffices to produce $h \in F$ such that $\|h\|_F < \epsilon$ and for some $\alpha > 0$, diam $S(f + g + h; \alpha) < 1/n$. By hypothesis, F contains a bump function b. Without loss of generality, $\|b\|_F < \epsilon$. By hypothesis (2), we can assume $b(0) \ne 0$ and therefore that $b(0) > 0$. Moreover, by hypothesis (3), we can assume that supp$(b) \subset B(0; 1/2n)$. Let $\alpha = b(0)/2$ and choose $x_0 \in E$ such that

$$(f + g)(x_0) < \inf_E (f + g) + b(0)/2.$$

Define h on E by $h(x) = -b(x - x_0)$; by hypothesis (2), $h \in F$ and $\|h\|_F = \|b\|_F < \epsilon$ and $h(x_0) = -b(0)$. To show that diam $S(f + g + h; \alpha) < 1/n$, it suffices to show that this set is contained in the ball $B(x_0; 1/2n)$, that is, if $\|x - x_0\| > 1/2n$, then $x \notin S(f + g + h; \alpha)$, the latter being equivalent to

$$(f + g + h)(x) > \inf_E (f + g + h) + \alpha.$$

Now, supp$(h) \subset B(x_0; 1/2n)$, so $h(x) = 0$ if $\|x - x_0\| > 1/2n$ hence

$$(f + g + h)(x) = (f + g)(x) \ge \inf_E (f + g) > (f + g)(x_0) - \alpha =$$

$$= (f + g + h)(x_0) + b(0) - b(0)/2 \ge \inf_E (f + g + h) + \alpha,$$

as was to be shown. Suppose that $g \in \bigcap U_n$; we want to show $g \in G$, that is, $f + g$ attains a strong minimum on E. First, for all n there exists $\alpha_n > 0$

such that $\operatorname{diam} S(f+g;\alpha_n) \leq 1/n$ and hence there exists a unique point $x_0 \in \bigcap S(f+g;\alpha_n)$. Suppose that $\{y_k\} \subset E$ and that $(f+g)(y_k) \to \inf_E(f+g)$. Given $n > 0$ there exists k_0 such that $(f+g)(y_k) \leq \inf_E(f+g) + \alpha_n$ for all $k \geq k_0$, therefore $y_k \in S(f+g;\alpha_n)$ for all $k \geq k_0$ and hence $\|y_k - x_0\| \leq \operatorname{diam} S(f+g;\alpha_n) \leq 1/n$ if $k \geq k_0$. Thus, $y_k \to x_0$ and therefore $g \in G$. A simple proof by contradiction show that $G \subset \bigcap U_n$, and completes the proof.

The first corollary to this theorem is a version of Ekeland's variational principle (Lemma 3.13). Whereas the latter produces a perturbation (by a small multiple of a translate of the norm) which attains a strict minimum, this corollary produces a similar perturbation which has a *strong* minimum. It does not, however, yield any control over the location of the minimum point.

Before proceeding, we should illustrate the difference between a strong minimum and strict minimum for such perturbations. Consider, for example, any nonreflexive Banach space E; by James' theorem [Di] there exists a continuous linear functional f of norm one on E which does not attain its norm; that is $f(x) < 1$ whenever $x \in E$, $\|x\| \leq 1$. [Concrete example: $f(x) = \Sigma 2^{-n} x_n$ on $E = c_0$.] Thus, if $x \neq 0$, then $f(-x/\|x\|) < 1$, that is $f(x) + \|x\| > 0$, which means that $f + \|\cdot\|$ attains a strict minimum at 0. On the other hand, since $\|f\| = 1$, there exists a sequence $\{x_n\} \subset E$, $\|x_n\| = 1$, such that $f(x_n) \to -1$ and hence $f(x_n) + \|x_n\| = f(x_n) + 1 \to 0$, although $x_n \not\to 0$.

Corollary 4.6. *Suppose that $f: E \to (-\infty, \infty]$ is proper, lower semicontinuous and bounded below. Then for all $\epsilon > 0$ there exists $x_0 \in E$ such that*

$$f(x_0) \leq \inf_E f + 2\epsilon$$

and the perturbed function $x \to f(x) + \epsilon\|x - x_0\|$ attains a strong minimum at x_0.

Proof. Let F be the space of all bounded real-valued Lipschitz continuous functions g on E with $\|g\|_F = \|g\|_\infty + \|g\|_{\text{Lip}}$, where

$$\|g\|_{\text{Lip}} = \sup\left\{\frac{|g(x) - g(y)|}{\|x - y\|} : x, y \in E, \ x \neq y\right\}.$$

It is straightforward to prove that F is a Banach space which satisfies hypotheses (1) through (3) of Theorem 4.5. To verify hypothesis (4), one can apply the construction in Prop. 4.3 to the norm in E to produce a bounded Lipschitzian bump function. Thus, there exists $g \in F$ such that $\|g\|_F < \epsilon$ and $f + g$ attains a strong minimum at some point $x_0 \in E$. Hence, for all $x \in E$,

$$|g(x)| < \epsilon, \quad |g(x) - g(x_0)| \leq \epsilon\|x - x_0\| \quad \text{and} \quad (f+g)(x) \geq (f+g)(x_0).$$

It follows that for all x we have $f(x) \geq f(x_0) + g(x_0) - g(x) \geq f(x_0) - \epsilon\|x - x_0\|$; that is, $x \to f(x) + \epsilon\|x - x_0\|$ attains its minimum at x_0. Also,

$$f(x) \geq f(x_0) + g(x_0) - g(x) \geq f(x_0) - 2\epsilon$$

so $\inf_E f \geq f(x_0) - 2\epsilon$. To see that x_0 is a strong minimum, suppose that $f(y_n) - g(x_0) + \epsilon\|y_n - x_0\| \to 0$; since

$$f(y_n) - f(x_0) + \epsilon\|y_n - x_0\| \geq (f+g)(y_n) - (f+g)(x_0),$$

we must have $y_n \to x_0$.

Our next application of Theorem 4.5 uses a different choice for the Banach space F.

Definition 4.7. Let D_β denote the linear space of all bounded Lipschitz continuous real-valued functions $g: E \to \mathbf{R}$ having bounded β-derivative $\nabla_\beta g$, provided with the norm

$$|||g||| = \|g\|_\infty + \|\nabla_\beta g\|_\infty$$

$$\equiv \sup\{|g(x)|: x \in E\} + \sup\{\|\nabla_\beta g(x)\|: x \in E\}.$$

It is clear that $||| \cdot |||$ makes D_β into a normed linear space, which will be nontrivial whenever hypothesis (H_β) is satisfied. Completeness of this space is obvious, until someone insists on the details. They are given in the following proposition.

Proposition 4.8. *The space* $(D_\beta, ||| \cdot |||)$ *is complete.*

Proof. For simplicity, if $g \in D_\beta$, we will write g' in place of $\nabla_\beta g$. By the mean value theorem [Fl], for any $f \in D_\beta$ and $x, y \in E$ we have $|g(x) - g(y)| \leq \|g'\|_\infty \cdot \|x - y\|$. Suppose, now, that $\{g_n\}$ is a $||| \cdot |||$-Cauchy sequence in D_β. Then both $\{g_n\}$ and $\{g'_n\}$ are uniformly Cauchy, so there exists $g: E \to \mathbf{R}$ and $h: E \to E^*$ such that $\{g_n\}$ converges uniformly to g and $\{g'_n\}$ converges uniformly to h. The boundedness of the sequence $\{g'_n\}$ implies that there is a bound on the Lipschitz constants for $\{g_n\}$, hence g is Lipschitzian. It remains to show that g is β-differentiable and that $g' = h$, that is, that for each $x \in E$ and $S \in \beta$, the difference quotients

$$\frac{g(x + ty) - g(x)}{t}$$

converge as $t \to 0^+$ to $\langle h(x), y \rangle$, uniformly for $y \in S$. To this end, fix $x \in E$ and $S \in \beta$ and let $M = \sup\{\|y\|: y \in S\}$. Define, for $t \in \mathbf{R}, t \neq 0$ and $y \in S$,

$$\phi_n(t, y) = \frac{g_n(x + ty) - g_n(x)}{t}$$

while $\phi_n(0, y) = \langle g'_n(x), y \rangle$. Given $\epsilon > 0$, the Cauchy property implies that there exists n such that $\|g'_m - g'_n\|_\infty < \epsilon$ provided $m \geq n$. The mean value theorem cited above shows that

$$|\phi_m(t, y) - \phi_n(t, y)| \leq \|g'_m - g'_n\|_\infty \cdot \|y\| \leq \epsilon M \tag{4.2}$$

for $m \geq n$, $t \in \mathbf{R}$ and $y \in S$. Since g_n is β-differentiable, there is a $\delta > 0$ such that $|\phi_n(t,y) - \phi_n(0,y) \leq \epsilon$ for all $y \in S$ and $0 < t < \delta$. Combined with (4.2), this shows that if $m \geq n$, $0 < t < \delta$ and $y \in S$, then

$$|\phi_m(t,y) - \phi_n(0,y)| \leq (1 + M)\epsilon. \tag{4.3}$$

By hypothesis, if $t \neq 0$, then

$$\phi_n(t,y) \to \phi(t,y) \equiv \frac{g(x+ty) - g(x)}{t},$$

while $\langle g'_n(x), y \rangle \to \langle h(x), y \rangle$. Let $m \to \infty$ in (4.3) to get $|\phi(t,y) - \phi_n(0,y)| \leq (1 + M)\epsilon$. Let $m \to \infty$ for $t = 0$ in (4.2) to get $|\langle g(x), y \rangle - \phi_n(0,y)| \leq \epsilon M$. From the last two inequalities

$$|\frac{h(x+ty) - h(x)}{t} - \langle g(x), y \rangle| \leq (1 + 2M)\epsilon$$

for all $0 < t < \delta$ and $y \in S$, which completes the proof.

The next corollary to Theorem 4.5 (which uses $F = D_\beta$) seems startling until one recalls that the intersection of a countable collection of dense G_δ subsets of a Banach space is itself a dense G_δ.

Corollary 4.9. *Suppose that E satisfies property (H_β) and that for each n, the function $f_n: E \to (-\infty, \infty]$ is proper, lower semicontinuous and bounded below. Then for any $\epsilon > 0$ one can choose $g \in D_\beta$ such that $\|g\|_\infty < \epsilon$, $\|\nabla_\beta g\|_\infty < \epsilon$ and each of the functions $f_n + g$ attains a strong minimum on E.*

Theorem 4.10. (DGZ Smooth Variational Principle) *Suppose that the Banach space E satisfies (H_β) and that f is a proper lower semicontinuous function on E which is bounded below. Then there exists a constant $a > 0$ (depending only on E) such that for all $0 < \epsilon < 1$ and for any $y_0 \in E$ such that $f(y_0) < \inf f + a\epsilon^2$, there exist $g \in D_\beta$ and $x_0 \in E$ such that*

a) $f + g$ has a strong minimum at x_0

b) $\|g\|_\infty < \epsilon$ and $\|\nabla_\beta g\|_\infty < \epsilon$

c) $\|x_0 - y_0\| \leq \epsilon$.

Proof. Let $b \in D_\beta$ be a bump function with $\mathrm{supp}(b) \subset B(0; 1)$ and $b(0) \geq 1$. By using an appropriate composition, we can assume that $0 \leq b \leq 1$ and $b(0) = 1$. [Indeed, let $\phi: \mathbf{R} \to [0,1]$ be a C^1 function which is monotonically nondecreasing and satisfies $\phi(0) = 0$ and $\phi(1) = 1$. Then $\phi \circ b$ has bounded derivative, is 0 outside $B(0; 1)$ and $\phi(b(0)) = 1$.] Define

$$M = \max\{\|\nabla_\beta b\|_\infty, 1\}, \qquad a = \frac{1}{4M},$$

and

$$h(x) = f(x) - 2a\epsilon^2 b(\frac{x - y_0}{\epsilon}), \qquad x \in E.$$

Note that $h\colon E \to (-\infty, \infty]$ is lower semicontinuous and bounded below. Thus, by applying Theorem 4.5 to $F = D_\beta$, there exists $k \in D_\beta$ with $\|k\|_\infty < a\epsilon^2/2$ and $\|\nabla_\beta k\|_\infty < \epsilon/2$ such that $h + k$ attains a strong minimum on E at x_0, say. We show first that $\|x_0 - y_0\| \leq \epsilon$. Suppose not; then $\|x_0 - y_0\| > \epsilon$, that is, $\frac{x_0 - y_0}{\epsilon} \notin B(0; 1)$ hence $b(\frac{x_0 - y_0}{\epsilon}) = 0$ so $h(x_0) = f(x_0) \geq \inf_E f$. Also,

$$h(y_0) = f(y_0) - 2a\epsilon^2 < \inf_E f - a\epsilon^2.$$

But $(h + k)(x_0) \leq (h + k)(y_0)$ so

$$h(x_0) \leq h(y_0) + k(y_0) - k(x_0) \leq h(y_0) + 2\|k\|_\infty < h(y_0) + a\epsilon^2,$$

hence $\inf_E f < h(y_0) + a\epsilon^2$, a contradiction. Next, let

$$g(x) = -2a\epsilon^2 b(\frac{x - y_0}{\epsilon}) + k(x),$$

so $g \in D_\beta$ and $f + g = h + k$ attains its minimum at x_0. Furthermore,

$$\|g\|_\infty \leq 2a\epsilon^2 \|b\|_\infty + \|k\|_\infty \leq 2a\epsilon^2 + a\epsilon^2/2 < \epsilon$$

(since $2a = \frac{1}{2M} < \frac{1}{2}$ hence $a/2 < 1/8$). Finally,

$$\|\nabla_\beta g\|_\infty \leq 2a\epsilon^2 \|\nabla_\beta b\|_\infty + \|\nabla_\beta k\|_\infty \leq 2a\epsilon^2 \dot{M} + \frac{\epsilon}{2} = \frac{1}{2}\epsilon^2 + \frac{\epsilon}{2} < \epsilon,$$

which completes the proof.

In order to apply this result, it is extremely useful to extend the notion of "subdifferential" from convex functions to arbitrary lower semicontinuous extended real-valued functions; in fact we will define the notion of "β-subdifferential" (as well as "β-superdifferential") for such functions.

Definition 4.11. Let $f\colon E \to (-\infty, \infty]$ be lower semicontinuous and suppose that $f(x)$ is finite. We say that f is *β-subdifferentiable* at x and that $x^* \in E^*$ is a *β-subdifferential* of f at x if, for each $\epsilon > 0$ and each set S in the bornology β, there exists $\delta > 0$ such that for $0 < t < \delta$

$$\langle x^*, y \rangle \leq \frac{f(x + ty) - f(x)}{t} + \epsilon \tag{4.4}$$

for all $y \in S$. We write $x^* \in \partial_\beta f(x)$.

It follows from this definition that if $\beta_1 \subset \beta_2$, then $\partial_{\beta_2} f(x) \subset \partial_{\beta_1} f(x)$. Now, we already have a definition of a subdifferential for a lower semicontinuous *convex* function f, so it is important to note that this new definition extends the old one. In fact, for such functions we have

$$\partial_\beta f(x) = \partial_G f(x) = \partial f(x)$$

for all choices of β. Indeed, by the previous remark, for any β we have $\partial_\beta f(x) \subset \partial_G f(x)$. On the other hand, if $x^* \epsilon \partial_G f(x)$, then for all $\epsilon > 0$ and $y \in E$ we have $\langle x^*, y \rangle \leq d^+ f(x)(y) + \epsilon$, so $x^* \leq d^+ f(x)$ and therefore $x^* \in \partial f(x)$. Finally, note that if $x^* \in \partial f(x)$, then the inequality (4.4) holds for all $\epsilon > 0$, $y \in E$ and $t > 0$; that is, for any β

$$\partial f(x) \subset \partial_\beta f(x).$$

We define β-*superdifferentials* by reversing the inequality (4.4) and replacing ϵ by $-\epsilon$; we denote the set of all these functionals by $\partial^\beta f(x)$. Note that

$$\partial^\beta(-f)(x) = -\partial_\beta f(x).$$

The next proposition exhibits the relations between the foregoing notions and β-differentiability. It also makes (in part (c)) the simple but key observation connecting the smooth variational principle with subdifferentiability.

Proposition 4.12. *(a) If f is lower semicontinuous, $f(x)$ is finite and both $\partial_\beta f(x)$ and $\partial^\beta f(x)$ are nonempty, then f is β-differentiable at x, and $\nabla_\beta f(x) = df(x)$.*

(b) Suppose that f is concave and continuous in a neighborhood of x. If $\partial_\beta f(x)$ is nonempty, then f is β-differentiable at x and $\partial_\beta f(x) = \{\nabla_\beta f(x)\}$.

(c) If g is β-differentiable at x and $f+g$ attains a minimum at x, then f is β-subdifferentiable at x (that is, $\partial_\beta f(x) \neq \emptyset$).

Proof. (a) Suppose that $x_1^* \in \partial^\beta f(x)$ and $x_2^* \in \partial_\beta f(x)$; then for any y in E and for every $\epsilon > 0$ we must have $\langle x_1^*, y \rangle - \langle x_2^*, y \rangle \leq 2\epsilon$, so $x_1^* = x_2^*$. Denote the common value by x^*; it follows readily from the definitions that $x^* = \nabla_\beta f(x)$.

(b) Since f is concave and continuous, the convex function $-f$ has a nonempty subdifferential and $\emptyset \neq \partial(-f)(x) = \partial_\beta(-f)(x) = -\partial^\beta f(x)$ (by the remarks following Definition 4.11). The hypothesis that $\partial_\beta f(x)$ be nonempty therefore implies that f is β-differentiable at x, by part (a).

(c) Note first that for all $u \in E$, $(-g)(u) - (-g)(x) \leq f(u) - f(x)$. Let $x^* = \nabla_\beta(-g)(x)$; then given $S \in \beta$ and $\epsilon > 0$, there exists $\delta > 0$ such that $0 < t < \delta$ and $y \in S$ imply that

$$-\epsilon \leq \frac{(-g)(x+ty) - (-g)(x)}{t} - \langle x^*, y \rangle \leq \frac{f(x+ty) - f(x)}{t} - \langle x^*, y \rangle,$$

which completes the proof.

Corollary 4.13. *If E has the property (H_β) and $f: E \to (-\infty, \infty]$ is a proper lower semicontinuous function, then*

$$\{x \in \mathrm{dom}(f): f \text{ is } \beta - \text{subdifferentiable at } x\}$$

is dense in $\mathrm{dom}(f)$.

Proof. Suppose that $f(x_0) < \infty$ and $\epsilon > 0$. There exists η such that $0 < \eta < \epsilon$ and $f(x) > f(x_0) - 1$ if $\|x - x_0\| \leq \eta$. Define the proper lower semicontinuous function $g: E \to \mathbf{R} \cup \{\infty\}$ by $g(x) = f(x)$ if $\|x - x_0\| \leq \eta$ while $g(x) = \infty$ otherwise, so g is bounded below. From property (H_β) it follows that there exists $\phi \in D_\beta$ such that $\phi(x_0) = 0$ and $\phi \equiv 2$ in $E \backslash B(x_0; \eta/2)$. Thus, we have $(g + \phi)(x_0) = f(x_0)$ and $(g + \phi) > f(x_0)$ in $E \backslash B(x_0; \eta/2)$. By Cor. 4.9, there exists $h \in D_\beta$ such that $\|h\|_\infty < 1/4$ and $g + \phi + h$ attains its minimum at some point x_1. It follows that $(g + \phi + h)(x) > f(x_0) + 3/4$ for $x \notin B(x_0; \eta/2)$ while $(g + \phi + h)(x_0) < f(x_0) + 1/4$. Necessarily, then, $x_1 \in B(x_0; \eta/2)$. Now, $f = g$ in $B(x_0; \eta/2)$ and $g + (\phi + h)$ attains its minimum at x_1; by Prop. 4.12(c), f is β-subdifferentiable at x_1.

Corollary 4.14. *Suppose that E has the property (H_β), that C is a nonempty open convex subset of E and $f: C \to \mathbf{R}$ is continuous and convex; then f is β-differentiable on a dense subset of C.*

Proof. Choose an open ball $B \subset C$ such that $\overline{B} \subset C$. Let $g = -f$ in $\overline{B}$ while $g = \infty$ in $E \backslash \overline{B}$, so g is proper and lower semicontinuous and hence there exists a dense subset of $\overline{B}$ at each point of which g is β-subdifferentiable. In particular, there exists $x \in B$ such that g is β-subdifferentiable at x. Since g is concave and continuous in B, by Prop. 4.12(b), g is β-differentiable at x, therefore f is β-differentiable at x.

This corollary does not say that the dense set of points of β-differentiability is a G_δ, but as we have seen (Prop. 1.25), this is automatic for Fréchet differentiability, hence we obtain the following corollary.

Corollary 4.15. *If a Banach space admits a Fréchet differentiable bump function with bounded derivative, then it is an Asplund space.*

This corollary is a slightly weaker form of the original theorem by I. Ekeland and G. Lebourg [Ek-L], who proved it without assuming boundedness of the derivative. By Prop. 4.3, their result (or the one above) solved half of a long-standing open problem, whether the existence of an equivalent Fréchet differentiable norm implies the Asplund property. The converse remained open for decades until R. Haydon's ingenious example [Ha₁]: *There exists a scattered compact Hausdorff space X with the property that $C(X)$ is an Asplund space but does not even admit an equivalent Gâteaux differentiable norm.* It had long been known that the converse *is* valid for separable E; more generally, Fabian [Fa₁] has shown that this is true if E is assumed to be WCG. (He assumes the formally weaker hypothesis that E be "weakly countably determined", but then shows that these are equivalent notions in Asplund spaces.)

Corollary 4.16. *If a Banach space E admits an equivalent Gâteaux differentiable norm (or, more generally, if it satisfies property (H_G)), then for every* *nonempty open convex subset D of E and continuous convex function f on D there exists a dense set of points in D where f is Gâteaux differentiable.*

This corollary came close to completely solving another long-standing problem: *If a Banach space admits an equivalent Gâteaux differentiable norm, is it necessarily a weak Asplund space?* (As was mentioned in Section 2, there exist continuous convex functions whose set G of points of Gâteaux differentiability is dense but G does not contain a dense G_δ subset.) Spaces which satisfy the conclusion of the Corollary 4.16 are called "Gâteaux Differentiability Spaces"; see Section 6. The fact that the existence of a Gâteaux differentiable norm *does* imply that the space is a weak Asplund space was proved much later by D. Preiss [P-P-N]; we will pursue this later in this section. Corollary 4.16 is sufficient to yield some negative results concerning smooth renormings.

Proposition 4.17. *The spaces ℓ^∞, $L^\infty[0,1]$ and $\ell^1(\Gamma)$ (Γ uncountable) do not admit equivalent Gâteaux differentiable norms.*

Proof. We showed in Example 1.4(b) that the norm in $\ell^1(\Gamma)$ is nowhere Gâteaux differentiable, and it is an interesting exercise (below) to prove the same fact about $L^\infty[0,1]$. Also, in Example 1.21 it was shown that $p(x) = \limsup |x_n|$ defines a continuous seminorm on ℓ^∞ which is nowhere Gâteaux differentiable.

4.18 Exercises.

(a) Prove that the essential supremum norm on the Banach space $L^\infty[0,1]$ is nowhere Gâteaux differentiable.

(b) Prove that any separable Banach space admits an equivalent strictly convex (rotund) norm. (Hint: Use a dual version of the construction in Exercise 2.40(b).)

(c) Show that there exists a strictly convex space whose dual norm is not smooth. (Hint: Look at the separable space ℓ^1 and Prop. 4.17).

We mentioned at the beginning of this section that Borwein and Preiss [Bor-P] were the first to establish a smooth variational principle. We will describe their result here and refer the reader to [Bor-P] or [Ph₄] for the proofs.

Definition 4.19 Let Θ denote the family of all real-valued functions θ on E of the form

$$\theta(x) = (1/2)\Sigma_{n=1}^\infty \mu_n \|x - v_n\|^2,$$

where $\Sigma\mu_n = 1$, $\mu_n \geq 0$ and $\{v_n\}$ converges in norm to some $v \in E$. Thus, each θ is a possibly infinite convex combination of squares of translates of the norm, with the translates themselves converging. It can be shown that if the norm in E is β-differentiable (at nonzero points), then each function in Θ is everywhere β-differentiable.

Theorem 4.20 (Borwein-Preiss). *Suppose that $g: E \to (-\infty, \infty]$ is lower semicontinuous and bounded below, that $\epsilon > 0$ and that $\lambda > 0$. Assume further that x_0 satisfies*

(a) $$g(x_0) < \inf_E g + \epsilon.$$

Then there exist $\theta \in \Theta$ and $v \in E$ such that

(b) $$g + (2\epsilon/\lambda^2)\theta \text{ attains its minimum on } E \text{ at } v$$

while

(c) $$\|x_0 - v\| < \lambda$$

and

(d) $$g(v) < \inf_E g + \epsilon.$$

Moreover, if E has a β-smooth norm, then θ is β-differentiable and

(e) $$\partial_\beta g(v) \cap (2\epsilon/\lambda)B^* \neq \emptyset,$$

where B^* is the dual unit ball.

Figure 4.1 below illustrates (a), (b), (c) and (d), where (b) is written as $g(x) \geq g(v) + (2\epsilon/\lambda^2)[\theta(v) - \theta(x)]$:

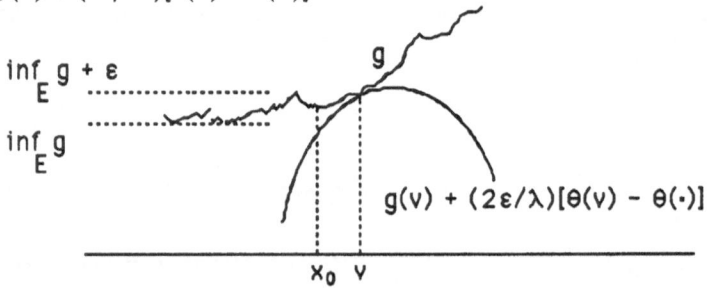

Fig. 4.1

One can apply this theorem to prove Cor. 4.13 (using the same arguments). While this result is more difficult to prove than Theorem 4.10, it has the advantage of giving (in conclusions (d) and (e)) additional information about the minimizing point of the perturbation of g.

The Borwein-Preiss theorem was motivated by the following deep theorem due to Preiss [Pr]. The proof is long and not easy; we content ourselves with merely giving the statement.

Theorem 4.21 (Preiss). *Any locally Lipschitzian real-valued function on an Asplund space is Fréchet differentiable at the points of a dense set.*

We return now to the fact that the existence of an equivalent Gâteaux differentiable norm on E implies that it is a weak Asplund space; more generally, it implies that maximal monotone operators are generically single-valued on E. The technique of proof (which, as we will see later, has application to yet more general results) is due to D. Preiss [P-P-N]. It also uses a topological result called the *Banach-Mazur game*, which we present next.

Definition 4.22. Suppose that X is a topological space and that S is any subset of X; we define a *game* (X, S) with players A and B, as follows. A *play* is a decreasing sequence $U_1 \supset V_1 \supset U_2 \supset V_2 \supset \ldots$ of nonempty open subsets of X which have been chosen alternately by A and B: Player A chooses U_1, B chooses V_1, A chooses U_2,... etc. A *strategy* for B is a sequence $f_B = \{f_n\}$ of maps f_n where, for each n, f_n is defined for $(U_1, V_1, U_2, V_2, \ldots, U_n)$ (the first $2n - 1$ elements of a play) and $f_n(U_1, V_1, U_2, V_2, \ldots, U_n)$ is a nonempty open subset of U_n. A play is *consistent with* $f_B = \{f_n\}$ provided $V_n = f_n(U_1, V_1, U_2, V_2, \ldots, U_n)$ for each n. We say that f_B is a *winning strategy* for B if $\bigcap V_n \subset S$ for every play consistent with f_B. (One can similarly define a winning strategy for A, but for our purposes, it is unnecessary.)

Recall that a set is said to be *residual* provided its complement is of first category.

We only need the "only if" portion of the following theorem; for a proof of the other half and related results, see [Ox].

Theorem 4.23 (Banach-Mazur). *Suppose that S is a subset of the topological space X and that A and B are players of the game (X, S). There exists a winning strategy for B if and only if S is residual in X.*

Proof. Suppose there exists a winning strategy $f_B = \{f_n\}$ for B. By an *f-chain of order n* we mean a nested sequence

$$U_1 \supset V_1 \supset U_2 \supset V_2 \supset \ldots \supset U_n \supset V_n$$

of nonempty open sets such that $V_i = f_i(U_1, V_1, \ldots U_i)$, $i = 1, 2, \ldots, n$. An f-chain of order $n + k$ is a *continuation* of one of order n if the first $2n$ terms of each are the same. We partially order the f-chains by continuation. Among all f-chains or order 1, let $\mathcal{F}_1$ be a maximal family with the property that any two distinct f-chains in $\mathcal{F}_1$ have disjoint smallest elements. This exists, by a straightforward application of Zorn's lemma. Since B has a strategy, the union W_1 of all the V_1's belonging to f-chains in $\mathcal{F}_1$ is dense in X. We now proceed by induction: Suppose that a family $\mathcal{F}_n$ of f-chains of order n has been defined so that the corresponding V_n's are pairwise disjoint and their union W_n is dense in X. Among the f-chains of order $n + 1$ which are continuations of members of $\mathcal{F}_n$, use Zorn's lemma to produce a maximal family $\mathcal{F}_{n+1}$ with the property that all the corresponding V_{n+1}'s are pairwise disjoint. The union W_{n+1} of the V_{n+1}'s is dense, by maximality and the existence of a strategy for B. Having found $\mathcal{F}_n$ for all n, let $W = \bigcap W_n$. For each $x \in W$, there exists a unique sequence $\{C_n\}$ of f-chains such that $C_n \in \mathcal{F}_n$ and x is in the corresponding V_n, for all n. Now, these f-chains C_n are linearly ordered by continuation, so we have $x \in \bigcap V_n$. This sequence is consistent with the strategy f_B by definition of f-chain; since B has a winning strategy, $x \in S$. Thus, we have shown that $W \equiv \bigcap W_n \subset S$ and hence $X \backslash S \subset \bigcup(X \backslash W_n)$ is of first category.

Our first application of the Banach-Mazur game is to a proof of the following basic theorem due to Asplund [Asp].

Theorem 4.24. *If E is a weak Asplund space and $T: E \to F$ is continuous, linear and onto the Banach space F, then F is a weak Asplund space.*

Proof. Suppose that $D \subset F$ is open and convex and that f is a continuous convex function on D. Then $D_1 = T^{-1}(D)$ is open and convex in E and $f_1 = f \circ T$ is continuous and convex on D_1. By hypothesis, there exists a dense G_δ subset $G_1 \subset D_1$ such that $df_1(x)$ exists for all $x \in G_1$. Let $G = T(G_1)$; then G is dense in D and $df(x)$ exists for all $x \in G$. Indeed, since T is onto, its adjoint T^* is one-one, and it is simple to see that $T^*(\partial f(Tx)) \subset \partial f_1(x)$ for all $x \in D_1$. In general, there is no reason to assume that G is a G_δ set, but the fact that it *contains* a dense G_δ set follows from the following corollary to Theorem 4.23.

Lemma 4.25. *Let M be a complete metric space, X a Hausdorff space and $f: M \to X$ a continuous, open surjective mapping. If $\{G_n\}$ is a sequence of dense open subsets of M, then the image $f(G)$ of $G = \bigcap G_n$ is residual in X.*

Proof. Let $S = f(G)$ and suppose that A and B play the Banach-Mazur game (X, S). The following reasoning will show that B has a winning strategy. Suppose that A has chosen the nonempty open subset U_1 of X. Since G_1 is dense and open in M, there exists an open metric ball $B_1 = B(x_1, r_1)$ such that $0 < r_1 < 1$ and $\overline{B}_1 \subset f^{-1}(U_1) \bigcap G_1$, so player B chooses $V_1 = f(B_1)$. Player A chooses some nonempty open subset $U_2 \subset V_1$. The set $B_1 \bigcap f^{-1}(U_2) \bigcap G_2$ is nonempty and open, so there exists $B_2 = B(x_2, r_2)$ with $0 < r_2 < 1/2$ and $\overline{B}_2 \subset B_1 \bigcap f^{-1}(U_2) \bigcap G_2$; player B takes $V_2 = f(B_2)$. Continuing in this way, for any sequence $U_1 \supset V_1 \supset U_2 \supset V_2 \supset \dots \supset U_n$ player B lets $V_n = f(B_n)$, where $B_n = B(x_n, r_n) \subset B_{n-1}$, $0 < r_n < 1/n$ and $\overline{B}_n \subset f^{-1}(U_n) \bigcap G_n$. These choices define a winning strategy for B, that is, $\bigcap V_n \subset f(G)$. Indeed, by completeness of M, $\bigcap \overline{B}_n$ is a single point, say $\{x_0\}$. If $y \in \bigcap V_n$, then for each n there exists $z_n \in B_n$ such that $y = f(z_n) \to f(x_0)$ and $\{x_0\} = \bigcap B_n \subset \bigcap G_n = G$ hence $y = f(x_0) \in f(G)$.

We next consider maximal monotone operators on Gâteaux smooth Banach spaces.

Definition 4.26. Let E be a Banach space, T a maximal monotone operator on E and $D = \text{int } D(T)$. Define

$$\sigma_T(x,y) = \sup \{\langle x^*, y\rangle \colon x^* \in T(x)\}, \quad x \in D, \ y \in E.$$

(We will usually write $\sigma(x,y)$ instead of $\sigma_T(x,y)$.)

If A is any subset of E^* and $y \in E$, we will, for simplicity, write $\{\langle A, y\rangle\}$ for the set of real numbers $\{\langle x^*, y\rangle \colon x^* \in A\}$.

We will also need a notion for monotone operators analogous to the situation when the directional derivative $df(x)(y)$ of f exists (at the point $x \in D$ in the direction $0 \neq y \in E$). Recall that this will be the case if and only if $d^+f(x)(-y) = -d^+f(x)(y)$. From Prop. 2.24 we know that $d^+f(x)(y) = \sup\{\langle x^*, y\rangle \colon x^* \in \partial f(x)\}$, so the equality above is equivalent to

$$\sup \{\langle \partial f(x), y\rangle\} = \inf \{\langle \partial f(x), y\rangle\},$$

that is, this set of real numbers is actually a singleton. This motivates the following definition.

Definition 4.27. For any $y \in E$ and T as above, let yT denote the set-valued mapping from E into the real line defined by

$$(yT)(x) = \{\langle x^*, y\rangle \colon x^* \in T(x)\}.$$

Our substitute for saying that $df(x)(y)$ exists will be the assertion that $yT(x)$ *is a singleton.*

Definition 4.28. The following notation will be convenient: If $A \subset E^*$, $y \in E$ and α is a real number, the assertion that $\langle A, y\rangle > \alpha$ means that $\langle x^*, y\rangle > \alpha$ for each $x^* \in A$.

Proposition 4.29. *Let T and D be as above. Then*

(i) For each $x \in D$, the real-valued function $y \to \sigma_T(x,y)$ is subadditive and positive homogeneous and, for any $\lambda > 0$, $\sigma_{\lambda T}(x,y) = \lambda \sigma_T(x,y)$.

(ii) For each $x \in D(T)$,

$$sup\{\sigma(x,y)\colon \|y\| = 1\} = sup\{\sigma(x,y)\colon \|y\| \leq 1\} = sup\{\|x^*\|\colon x^* \in T(x)\}.$$

(iii) $(yT)(x)$ is a singleton if and only if $\sigma(x, -y) = -\sigma(x,y)$.

(iv) If $x_0 \in D$, $y \in E$ and $(yT)(x_0)$ is a singleton (say equal to $\{\alpha\}$), then for all $\epsilon > 0$ there exists a neighbohood U of x_0 in D such that $\langle T(U), y\rangle > \alpha - \epsilon$.

(v) Fixing $x \in D$ and $y \neq 0$, letting $I = \{t \in R \colon x + ty \in D\}$ and defining

$$f(t) = \sigma(x + ty, y), \quad t \in I,$$

the function f is monotone nondecreasing on I (and hence is continuous at all but countably many points of I). Moreover, if f is continuous at $t_0 \in I$, then $(yT)(x + t_0 y)$ is a singleton.

Proof. (i) and (ii) are immediate from the definitions.

(iii) If $(yT)(x)$ is a singleton, then so is $(-yT)(x)$, with $(-yT)(x) = -\sigma(x, y)$; that is, $\sigma(x, -y) = -\sigma(x, y)$. On the other hand, if $(yT)(x)$ is not a singleton, there exists $x^* \in T(x)$ such that $\langle x^*, y \rangle < \sigma(x, y)$ and hence $\sigma(x, -y) \geq \langle x^*, -y \rangle \geq -\sigma(x, y)$.

(iv) Given $\epsilon > 0$, let $W = \{y^* \in E^*: \langle y^*, y \rangle > \alpha - \epsilon\}$. This is a weak* open set containing $T(x_0)$, so by the norm-to-weak* upper semicontinuity of T at x_0, there exists an open neighborhood U of x_0 in D such that $T(x) \subset W$ for all $x \in U$, which was to be proved.

(v) Note that we are not asserting that the open set I is an interval, but this does not affect our argument. Suppose that $t_1, t_2 \in I$ with $t_1 < t_2$; then for any $x_i^* \in T(x + t_i y)$, $i = 1, 2$, we have (by monotonicity)

$$0 \leq \langle x_1^* - x_2^*, (x + t_1 y) - (x + t_2 y) \rangle = (t_1 - t_2)\langle x_1^* - x_2^*, y \rangle,$$

hence $\langle x_1^*, y \rangle \leq \langle x_2^*, y \rangle$. This shows that

$$f(t_1) \equiv \sup\{\langle x^*, y \rangle: x^* \in T(x + t_1 y)\} \leq \inf\{\langle x^*, y \rangle: x^* \in T(x + t_2 y)\}; \quad (4.5)$$

in particular, $f(t_1) \leq f(t_2)$, so f is monotone. Suppose, now, that yT is not single-valued at $x + t_0 y$. Then

$$\alpha \equiv \inf \{\langle x^*, y \rangle: x^* \in T(x + t_0 y)\} < f(t_0).$$

Hence if $t \in I$, $t < t_0$, then (by (4.5)), $f(t) \leq \alpha < f(t_0)$, so f is not continuous at t_0.

The relationship between Gâteaux smoothness of the norm and single-valuedness of T is exhibited by the following lemma.

Lemma 4.30. *Suppose that $x_0 \in D$ and that $y_0 \in E$, $\|y_0\| = 1$, is such that $y_0 T$ is single-valued at x_0, with value α. If*

$$\sup\{\sigma_T(x_0, y): \|y\| = 1\} \leq \alpha,$$

then

$$T(x_0) \subset \alpha \cdot \partial\| \cdot \|(y_0).$$

In particular, if the norm is Gâteaux differentiable at y_0, then $T(x_0)$ is a singleton.

Proof. Suppose that $x^* \in T(x_0)$; then from Prop. 4.29 (ii), we see that $\alpha \geq 0$ and $\|x^*\| \leq \alpha$. Since $(y_0 T)(x_0)$ is a singleton, it follows that $\langle x^*, y_0 \rangle = \alpha$ and so $x^* \in \alpha \cdot \partial\| \cdot \|(y_0)$.

Theorem 4.31. *Suppose that E admits an equivalent Gâteaux smooth norm, that T is a maximal monotone operator on E and that*

$$D \equiv int \ \{x \in E: \ T(x) \neq \emptyset\}$$

is nonempty. Then there exists a dense G_δ subset $G \subset D$ such that $T(x)$ is a singleton for every $x \in G$.

Proof. Let $p_1 = \|\cdot\|$ denote an equivalent smooth norm on E. Choose sequences of positive numbers $1/2 > \ \epsilon_1 > \ \epsilon_2 > \cdots$ and $\beta_1 > \beta_2 > \beta_3 > \cdots$ such that

$$\epsilon_k \to 0, \quad \Sigma\beta_k^2 < 3 \quad \text{and} \quad \Sigma\sqrt{\epsilon_k}/\beta_k < \infty.$$

To use the Banach-Mazur game we let D be our Hausdorff space and S the set of points in D where T is single-valued. After player A chooses a nonempty open subset U of D, player B's first choice will always be an open subset of U in which T is bounded. Thus, we may always assume that player A's first choice was an open nonempty set U_1 in which T is bounded. We may also assume that

$$\sup \ \{p_1^*(x^*): \ x^* \in T[U_1]\} > 0.$$

[Indeed, if this supremum is 0, then B's strategy is obvious: Since T is single-valued (equal to 0) on the entire set U_1, she need only choose $V_k = U_k$ for each $k = 1, 2, 3, \ldots$ so that $\cap V_k = U_1 \subset S$.] Now, let S_1 be the unit sphere $\{x \in E: \ p_1(x) = 1\}$ defined by p_1 and define

$$s_1 = \sup \ \{\sigma(x,y): \ (x,y) \in U_1 \times S_1\}.$$

From Prop. 4.29 (ii), we can also write

$$s_1 = \sup \ \{p_1^*(x^*): \ x^* \in T(x) \text{ and } x \in U_1\};$$

by our earlier assumption, $s_1 > 0$. From part (v) of Prop. 4.29, for any $y \in E$ and $x \in U_1$, there exist points of the form $x + ty \in U_1$ with $t > 0$ such that $\sigma(x,y) \leq \sigma(x + ty, y)$ and yT is single-valued at $x + ty$. Thus

$$s_1 = \sup \ \{\sigma(x,y): \ (x,y) \in U_1 \times S_1 \text{ and } yT \text{ is single-valued at } x\}.$$

It follows that there exists $(x_1, y_1) \in U_1 \times S_1$ such that $y_1 T$ is single-valued at x_1 and such that $\langle T(x_1), y_1 \rangle > (1 - \epsilon_1)s_1$. By Prop. 4.29 (iv), there exists r_1, with $0 < r_1 < 1$, such that $B(x_1, 2r_1) \subset U_1$ and $\langle T[B(x_1; 2r_1)], y_1 \rangle > (1 - \epsilon_1)s_1$. Define $V_1 = B(x_1, r_1)$. For all $x \in E$ define

$$q_1(x) = \text{dist} \ (x, Rx_1) \equiv \inf \ \{\|x - \lambda x_1\|: \ \lambda \in R\}$$

and define a new norm p_2 on E by

$$p_2^2 = p_1^2 + \beta_1^2 q_1^2.$$

We now have s_1, y_1, x_1, p_2 and player B's open set V_1, so player A may choose any nonempty open subset $U_2 \subset V_1$. Using a similar strategy to respond to player A's choices $U_2, U_3, \ldots$ at every step, player B chooses $V_1, V_2, \ldots$ by constructing sequences of numbers $s_1, s_2, \ldots, s_k, \ldots$, norms $p_1, p_2, \ldots, p_k, \ldots$ (with dual norms p_k^*), spheres $S_1, S_2, \ldots, S_k = \{x \in E: p_k(x) = 1\}, \ldots$ containing the vectors $y_1, y_2, \ldots, y_k, \ldots$ respectively, points $x_1, x_2, \ldots, x_k, \ldots$ and positive radii $r_1, r_2, \ldots, r_k, \ldots$ such that $V_k = B(x_k, r_k)$,

$$p_k^2 = p_{k-1}^2 + \beta_{k-1}^2 \cdot q_{k-1}^2 \equiv p_1^2 + \sum_{j=1}^{k-1} \beta_j^2 q_j^2,$$

(where $q_k(x) = \inf\{\|x - \lambda x_k\|: \lambda \in R\}, \quad x \in E$),

$$s_k = \sup\{\sigma(x,y): (x,y) \in U_k \times S_k\}$$
$$= \sup\{p_k^*(x^*): x^* \in T(U_k)\},$$

$y_k T$ is single-valued at x_k and

$$\langle T(x), y_k \rangle > (1 - \epsilon_k)s_k \text{ for } x \in B(x_k, 2r_k) \subset U_k.$$

Since $B(x_k, 2r_k) \subset U_k \subset V_{k-1} = B(x_{k-1}, r_{k-1})$, we have

$$r_k \leq (1/2)r_{k-1} \leq (1/2^2)r_{k-2} \leq \ldots \leq (1/2^{k-1})r_1 < 1/2^{k-1}$$

as well as $\overline{V_k} \subset U_k$ for each k. Also, $p_k \geq p_{k-1}$ implies that

$$S_k \subset B_{k-1} = \{x \in E: p_{k-1}(x) \leq 1\},$$

so, using $U_k \subset U_{k-1}$ and Prop. 4.29(ii),

$$s_k \leq \sup\{\sigma(x,y): (x,y) \in U_{k-1} \times B_{k-1}\} =$$
$$= \sup\{\sigma(x,y): (x,y) \in U_{k-1} \times S_{k-1}\} = s_{k-1}.$$

Note that $p_k(y_{k-1}) = p_{k-1}(y_{k-1}) = 1$ and $x_k \in U_k \subset U_{k-1}$, so necessarily $(x_k, y_{k-1}) \in U_k \times S_k$ hence $s_k \geq \sigma(x_k, y_{k-1})$, while $x_k \in U_{k-1}$ implies that $\sigma(x_k, y_{k-1}) > (1 - \epsilon_{k-1})s_{k-1}$. Thus, $s_k > (1 - \epsilon_{k-1})s_{k-1}$. Since $s_1 > 0$ and $\epsilon_1 < 1$, this implies that $s_2 > 0$; by induction, $s_k > 0$ for all k. It follows that the decreasing sequence $\{s_k\}$ converges to some number $s_\infty \geq 0$. Also, since the diameters of the sets V_k converge to zero, the intersections of their closures consists of a single point, denoted by x_∞; necessarily, $x_\infty \in V_k$ for all k. In order to be able to apply Lemma 4.30 and complete the proof, we will show that

(i) the sequence of norms $\{p_k\}$ converges to an equivalent smooth norm p_∞ satisfying $p_1 \leq p_\infty \leq 2p_1$,

(ii) the sequence of vectors $\{y_k\}$ is convergent, with limit y_∞ satisfying $p_\infty(y_\infty) = 1$ and

(iii) $y_\infty T$ is single-valued at x_∞, with value s_∞.

Assuming that we have proved assertions (i), (ii) and (iii), we next prove the inequality in Lemma 4.30, taking, of course, x_∞ and y_∞ in place of x_0

and y_0, s_∞ in place of α and p_∞ for the norm; that is, we want to show that $\sigma(x_\infty, y) \leq s_\infty$ whenever $p_\infty(y) = 1$. For each k, we have $x_\infty \in U_k$ and $p_k(y)^{-1} y \in S_k$ and therefore

$$\sigma(x_\infty, y) = p_k(y) \cdot \sigma(x_\infty, p_k(y)^{-1} y) \leq s_k \cdot p_k(y).$$

This last term converges to $s_\infty \cdot p_\infty(y) = s_\infty$, so the inequality of Lemma 4.30 is satisfied and therefore $T(x_\infty) \subset s_\infty \cdot \partial p_\infty(y_\infty)$. Since p_∞ is smooth, $T(x_\infty)$ is a singleton; that is $\{x_\infty\} = \cap V_k \subset S$.

We have thus described a winning strategy for player B, so by Theorem 4.23, S contains a dense G_δ subset (the open set D being completely metrizable). It now remains to prove assertions (i), (ii) and (iii).

(i) Since $q_k \leq p_1$ for each k, we have

$$p_1^2 \leq p_k^2 = p_1^2 + \sum_{j=1}^{k-1} \beta_j^2 q_j^2 \leq (1 + \sum_{j=1}^{k-1} \beta_j^2) \cdot p_1^2 \leq 4 \cdot p_1^2,$$

hence the increasing sequence $\{p_k\}$ of equivalent norms converges uniformly on bounded sets to a norm p_∞ satisfying $p_1 \leq p_\infty \leq 2p_1$. To prove that p_∞ is Gâteaux smooth, note first that each of the functions q_k^2 is everywhere Gâteaux differentiable. In fact, if $q_k^2(y) = 0$, then it is easily seen that $\partial q_k^2(y) = \{0\}$. If $q_k^2(y) > 0$, then for some $\lambda_0 \in R$,

$$0 < q_k(y) \equiv \inf \{\|y - \lambda y_k\| : \lambda \in R\} = p_1(y - \lambda_0 y_k)$$

and hence, for any $t > 0$ and $u \in E$,

$$\begin{aligned} 0 &\leq t^{-1}[q_k(y + tu) + q_k(y - tu) - 2q_k(y)] \leq \\ &\leq t^{-1}[p_1(y - \lambda_0 y_k + tu) + p_1(y - \lambda_0 y_k - tu) - 2p_1(y - \lambda_0 y_k)]. \end{aligned} \quad (4.6)$$

Since Gâteaux differentiability is characterized by the fact that such difference quotients converge to zero (Exercise 1.24), Gâteaux differentiability of p_1 at points where it is positive implies that the last term in (4.6) converges to zero, hence so does the first term. It now follows by standard arguments that the infinite series defining p_∞^2 is everywhere Gâteaux differentiable, hence p_∞ is Gâteaux differentiable at nonzero points.

(ii) To show that $\{y_k\}$ converges, we will show that for all k,

$$\|y_{k+1} - y_k\| \leq 6\sqrt{\epsilon_k}/\beta_k;$$

since the series $\Sigma \sqrt{\epsilon_k}/\beta_k$ converges, this will suffice to show that $\{y_k\}$ is a Cauchy sequence, hence is convergent to y_∞, say. To obtain the estimate above, recall first that $s_{k+1} > (1 - \epsilon_k)s_k$ and $\epsilon_k > \epsilon_{k+1}$ so

$$\sigma(x_{k+1}, y_{k+1}) > (1 - \epsilon_{k+1})s_{k+1} > (1 - \epsilon_{k+1})(1 - \epsilon_k)s_k > (1 - 2\epsilon_k)s_k > 0,$$

while — since $(x_{k+1}, p_k(y_{k+1})^{-1} y_{k+1}) \in U_k \times S_k$ — we have

$$\sigma(x_{k+1}, y_{k+1}) = p_k(y_{k+1}) \cdot \sigma(x_{k+1}, p_k(y_{k+1})^{-1} y_{k+1}) \leq s_k \cdot p_k(y_{k+1}).$$

Thus $p_k(y_{k+1}) > 1 - 2\,\epsilon_k > 0$. By the definition of p_{k+1},

$$\beta_k^2 q_k^2(y_{k+1}) = p_{k+1}^2(y_{k+1}) - p_k^2(y_{k+1}) < 1 - (1 - 2\epsilon_k)^2 < 4\epsilon_k.$$

Consequently, dist $(y_{k+1}, Re_k) \equiv q_k(y_{k+1}) < 2\sqrt{\epsilon_k}/\beta_k$. We can write a nearest point in Re_k to y_{k+1} in the form $\lambda_k y_k$ for some $\lambda_k \in R$; that is

$$y_{k+1} = \lambda_k y_k + u_k, \text{ where } \|u_k\| = q_k(y_{k+1}) < 2\sqrt{\epsilon_k}/\beta_k.$$

Now, $p_{k+1} \leq p_\infty \leq 2p_1$, so $p_{k+1}(u_k) \leq 4\sqrt{\epsilon_k}/\beta_k$. Since, as noted earlier, $p_{k+1}(y_k) = p_k(y_k) = 1$, we have

$$|\lambda_k| = p_{k+1}(\lambda_k y_k) = p_{k+1}(y_{k+1} - u_k) \geq 1 - p_{k+1}(u_k) \geq 1 - 4\sqrt{\epsilon_k}/\beta_k.$$

Also,

$$|\lambda_k| \leq p_{k+1}(y_{k+1}) + p_{k+1}(-u_k) \leq 1 + 4\sqrt{\epsilon_k}/\beta_k.$$

We next show that if k is large enough so that $\epsilon_{k+1} + 4\sqrt{\epsilon_k}/\beta_k < 1$, then the constant λ_k is positive. Indeed, since $x_{k+1} \in V_k$ for all k, we must have $0 < (1 - \epsilon_k)s_k < \sigma(x_{k+1}, y_k)$. Choose an element $x_k^* \in T(x_{k+1})$ such that $\langle x_k^*, y_{k+1} \rangle > (1 - \epsilon_k)s_k$. (Note that by Prop. 4.29 (ii), $p_{k+1}^*(x_k^*) \leq s_{k+1}$.) Since $(y_{k+1}T)(x_{k+1})$ is a singleton, we must have

$$\langle x_k^*, y_{k+1} \rangle = \sigma(x_{k+1}, y_{k+1}) > (1 - \epsilon_{k+1})s_{k+1}.$$

It follows that

$$(1 - \epsilon_{k+1})s_{k+1} < \langle x_k^*, \lambda_k e_k + u_k \rangle = \lambda_k \langle x_k^*, y_k \rangle + \langle x_k^*, u_k \rangle \leq$$
$$\leq \lambda_k \langle x_k^*, y_k \rangle + s_{k+1} \cdot 4\sqrt{\epsilon_k}/\beta_k,$$

therefore $\lambda_k \langle x_k^*, y_k \rangle > (1 - \epsilon_{k+1} - 4\sqrt{\epsilon_k}/\beta_k)s_{k+1}$. By hypothesis, this last term is positive and since $\langle x_k^*, y_k \rangle > 0$ we conclude that $\lambda_k > 0$ for all sufficiently large k. It follows that our earlier estimates on $|\lambda_k|$ become $1 - 4\sqrt{\epsilon_k}/\beta_k \leq \lambda_k \leq 1 + 4\sqrt{\epsilon_k}/\beta_k$, that is, $|1 - \lambda_k| \leq 4\sqrt{\epsilon_k}/\beta_k$. Consequently, using the fact that $\|y_k\| \equiv p_1(y_k) \leq p_k(y_k) = 1$, we finally obtain the desired estimate:

$$\|y_{k+1} - y_k\| = \|(\lambda_k - 1)y_k + u_k\| \leq |\lambda_k - 1| \cdot \|y_k\| + \|u_k\| \leq$$
$$\leq |\lambda_k - 1| + \|u_k\| \leq 4\sqrt{\epsilon_k}/\beta_k + 2\sqrt{\epsilon_k}/\beta_k = 6\sqrt{\epsilon_k}/\beta_k.$$

The fact that $p_\infty(y_\infty) = 1$ is immediate from the fact that $p_k \to p_\infty$ uniformly on bounded sets, since $p_k(y_k) = 1$ and $y_k \to y_\infty$.

(iii) Finally, we show that $y_\infty T$ is single-valued at x_∞, with value s_∞. To that end, suppose that $x^* \in T(x_\infty)$. Since $x_\infty \in U_k$ for all k and since $1 = p_\infty(y_\infty) \geq p_k(y_\infty)$, the fact (proved earlier) that

$$s_k = \sup\{\sigma(x, y): x \in U_k, \quad p_k(y) \leq 1\}$$

implies that $s_k \geq \sigma(x_\infty, y_\infty) \geq \langle x^*, y_\infty \rangle$, hence $s_\infty \geq \langle x^*, y_\infty \rangle$. On the other hand, since $x^* \in T[U_1]$, we have $\|x^*\| \leq s_1$ and $x_\infty \in V_{k-1}$ for all k, so

$$\langle x^*, y_\infty \rangle \geq \langle x^*, y_k \rangle - s_1 \|y_k - e_\infty\| > (1 - \epsilon_k)s_k - s_1 \|y_k - y_\infty\|;$$

it follows that $\langle x^*, y_\infty \rangle = s_\infty$ and the proof is complete.

The utility of "smoothability" (the existence of an equivalent Gâteaux differentiable norm) in Theorem 4.31 raises anew the question of permanence properties for such spaces. It is easy to see that smoothability is inherited by subspaces and preserved under finite products. It does not satisfy the "three-space property"; this was shown by Talagrand [Ta$_2$] when he showed that the space $E_1 = C(K)$, with K the compact Hausdorff "two arrow space," does not admit an equivalent smooth norm, even though it contains the separable (hence weak Asplund) subspace $C[0,1]$, while the quotient space $E_1/C[0,1]$ is isometric to $c_0(0,1]$ (which is an Asplund space). Theorem 4.31 provides another proof of this result: If E_1 admitted an equivalent smooth norm, then it would be a weak Asplund space, but M. Čoban and P. S. Kenderov [Č-K] have shown that the set of points of Gâteaux differentiability of the supremum norm in E_1 does not contain a dense G_δ subset. As to permanence properties of weak Asplund spaces, the example E_1 described above shows that *weak Asplund spaces do not satisfy the three-space property.*

A natural question about monotone operators remains open: *If E is a weak Asplund space, is every maximal monotone operator on E generically single-valued?*

Remarks.

The alert reader will have noticed that we have not made full use of Theorem 4.10 (the DGZ smooth variational principle), since we give no applications to Hadamard or weak Hadamard differentiability. Of course, the former coincides Gâteaux differentiability for locally Lipschitzian functions, while the latter coincides with Fréchet differentiability in reflexive spaces. J. Borwein [Bor$_3$] and P. Ørno [Ør] have shown that for continuous convex functions, weak Hadamard differentiability coincides with Fréchet differentiability in any Banach space which does not contain a subspace isomorphic to ℓ_1. J. Borwein and S. Fitzpatrick [Bor-F$_2$] have shown that for a σ-finite measure μ, the space $L^1(\mu)$ admits an equivalent weak Hadamard differentiable norm, hence Theorem 4.10 applies to closed subspaces of such L^1 spaces. In [DGZ$_3$] the authors apply their smooth variational principle to generalize the kinds of results on differential equations in Banach spaces proved originally by M. Crandall and P.- L. Lions in [Cr-Li$_{1,2}$].

In the original version of Theorem 4.20, Borwein and Preiss utilized perturbation functions made up of sums of p-th powers of translates of the norm, $1 \leq p < \infty$; it seemed simpler to us to restrict attention to the case $p = 2$. Their paper lists several other bornologies, hence several other kinds of differentiability.

Theorem 4.24 (that the weak Asplund property is preserved by continuous linear surjective maps) was proved by Asplund [Asp], but his proof seems to assume that the image of a dense G_δ under such a mapping is a G_δ, something which even fails in $\mathbf{R}^2$.

This section has touched briefly on a number of results concerning Gâteaux or Fréchet smooth renormings of Banach spaces and their relationship to the existence of smooth bump functions. For a comprehensive treatment of such topics, see the forthcoming book by R. Deville, G. Godefroy and V. Zizler [D-G-Z$_2$]. The same authors have utilized the construction in the proof of Theorem 4.31 to generalize it to the following result [D-G-Z$_1$]: *If E admits a Gâteaux differentiable and Lipschitzian bump function, then E is a weak Asplund space.* A different approach to proving that Gâteaux smooth spaces are weak Asplund spaces (again using the construction in Theorem 4.31) has been obtained by N. Ribarska [Ri$_3$]. She avoids the Banach-Mazur game by applying her earlier results on "fragmentability" [Ri$_{1,2}$], a concept which plays a key role in the interplay between general topology and geometry of Banach spaces.

There is a natural and straightforward definition of "β-continuity" of a monotone operator T which reduces to β-differentiability of f when T is the subdifferential of the convex continuous function f. Another application of the construction in Theorem 4.31 has been made in [P-P-N] in order to prove that *if E admits an equivalent β-smooth norm, then the interior of the domain of any maximal monotone operator on E contains a dense G_δ set of points at which the operator is β-continuous.*

Section 5

5. Asplund spaces, the RNP and perturbed optimization.

In this section we are going to look at a duality relationship between Asplund spaces and spaces with the Radon-Nikodým property (RNP). Roughly speaking, a Banach space E or, more generally, a closed convex subset C of E, is said to have the Radon-Nikodým property if the classical Radon-Nikodým theorem (on the representation of absolutely continuous measures in terms of integrals) is valid for vector-valued measures whose "average range" is contained in C. This property has been characterized in purely geometrical terms (which is the basis of the definition we use below). For the extraordinarily wide range of connections between the RNP and various parts of integration theory, operator theory and convexity, one should read the 1977 survey by Diestel and Uhl [Di-U] and, for more recent results (1983) the comprehensive lecture notes by Bourgin [Bou].

Definition 5.1. (a) Recall, first, the definition of a *slice* $S(x^*, A, \alpha)$ of a nonempty subset A of a Banach space E: For $\alpha > 0$ and $x^* \in E^*$,

$$S(x^*, A, \alpha) = \{x \in A : \langle x^*, x \rangle > \sigma_A(x^*) - \alpha\}.$$

If A is a nonempty subset of E^*, we define a *weak* slice* analogously, with the functional coming from E rather than from E^{**}.

(b) We say that a nonempty subset A of E is *dentable* provided it admits slices of arbitrarily small diameter; that is, for every $\epsilon > 0$ there exists $x^* \in E^*$ and $\alpha > 0$ such that diam $S(x^*, A, \alpha) < \epsilon$.

(c) If A is a nonempty subset of E^*, we say that it is *weak*-dentable* provided it admits weak*-slices $S(x, A, \alpha)$ of arbitrarily small diameter.

Using this terminology, Theorem 2.31 can be reformulated to say that *a Banach space E is an Asplund space if and only if every nonempty bounded subset of E^* is weak* dentable.*

Definition 5.2. A subset A of a Banach space E is said to have the *Radon-Nikodým property* (RNP) if every nonempty bounded subset of A is dentable.

Since weak* dentable sets are obviously dentable, Theorem 2.31 implies that *if E is an Asplund space, then E* has the RNP*. One of the most striking results in this area is that the reverse implication is also valid. To that end, we need another definition.

Definition 5.3. *An infinite tree* in a Banach space E is a sequence $\{x_n\}$ such that $x_n = (1/2)(x_{2n} + x_{2n+1})$ for each n. An infinite tree such that $\|x_{2n} - x_{2n+1}\| \geq 2\delta$ for some $\delta > 0$ and all n is called an *infinite δ-tree*.

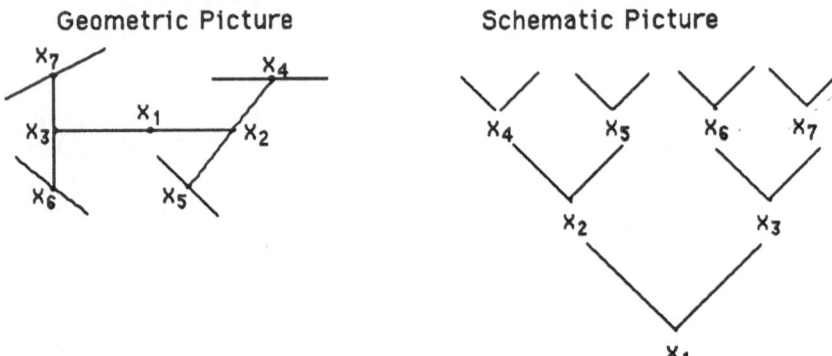

Fig. 5.1

Here is an obvious but important observation: *An infinite δ-tree cannot be dentable* (since every slice must have diameter at least 2δ). What we will do is use the characterization that E is an Asplund space (if and) only if every separable subspace of E has separable dual (Theorem 2.33) in order to produce a δ-tree in E^*. We need some preliminary results.

Lemma 5.4. *Assume that E is a Banach space which contains a separable subspace F such that F^* is nonseparable. Then there exists $\epsilon > 0$ and a nonempty subset A of the unit ball B^* of E^* such that every nonempty relatively weak* open subset of A has diameter greater than ϵ.*

Proof. Let $B(F^*)$ be the unit ball of the nonseparable space F^*. By the proof of Theorem 2.19, there exist $\epsilon > 0$ and an uncountable subset A_1 of $B(F^*)$ such that each point of A_1 is a weak* condensation point of A_1 and such that $\|x^* - y^*\| > \epsilon$ whenever x^*, y^* are distinct points of A_1. It follows that any nonempty relatively weak*-open subset of the weak* closure A_2 of A_1 contains at least two distinct points of A_1. The restriction map $R: E^* \to F^*$ maps B^* onto $B(F^*)$ and is weak*-to-weak* continuous. Let A be a minimal weak* compact subset of B^* such that $R(A) = A_2$. Thus, if U is a relatively weak* open subset of A, then the image A_3 under R of the weak* compact set $A \setminus U$ is a proper closed subset of A_2. Since $A_2 \setminus A_3$ is a relatively weak* open subset of A_2 it contains at least two distinct points of A_1 and hence there exist distinct points x^*, y^* in U with $\|x^* - y^*\| > \epsilon$.

Lemma 5.5. *Let $\{K_n\}$ be a sequence of nonempty compact convex subsets of a topological vector space such that $K_{2n} \cup K_{2n+1} \subset K_n$ for all n. Then there exists an infinite tree $\{x_n\}$ such that $x_n \in K_n$ for all n.*

Proof. Let $K = \Pi K_n$ be the (compact) cartesian product consisting of all $x = (x_n)$ such that $x_n \in K_n$ for each n. Define, for each n,

$$A_n = \{x = (x_k) \in K : x_n = (1/2)(x_{2n} + x_{2n+1})\}.$$

Each A_n is closed, hence compact, and a sequence $\{x_n\}$ corresponding to an element $x \in K$ will be an infinite tree if $x \in \cap A_n$, that is, we need only show that this intersection is nonempty. By compactness, it suffices to prove that $A_1 \cap A_2 \cap \cdots \cap A_k \neq \emptyset$ for each k. To this end, fix k and define x in K as follows: If $n > k$, let x_n be any element of K_n. From $n = k$ to $n = 1$, use induction: Suppose that, for $m = n+1, n+2, \ldots$, a point $x_m \in K_m$ has been chosen, and define $x_n = (1/2)(x_{2n} + x_{2n+1})$. Clearly, $x_n \in K_n$ since the latter is convex and $K_{2n} \cup K_{2n+1} \subset K_n$. The resulting element x is in $A_1 \cap A_2 \cap \cdots \cap A_k$, so the proof is complete.

Proposition 5.6. *Suppose that E is a Banach space and suppose that there exists a nonempty bounded set $A \subset E^*$ and an $\epsilon > 0$ such that diam $U > \epsilon$ whenever U is a nonempty relatively weak* open subset of A. Then its weak* closed convex hull w^*-cl co$A$ contains an infinite $\epsilon/2$-tree.*

Proof. We construct a sequence $\{U_n\}$ of nonempty relatively weak* open subsets of A and a sequence $\{x_n\}$ in E such that

(a) $\|x_n\| = 1$ for each n,

(b) $U_{2n} \cup U_{2n+1} \subset U_n$, $n = 1, 2, 3, \ldots$, and

(c) $x^* \in U_{2n}$ and $y^* \in U_{2n+1}$ imply that $\langle x^* - y^*, x_n \rangle \geq \epsilon$ for each n.

First, let $U_1 = A$. Suppose that, for some positive integer m, sets U_k have been defined for $1 \leq k < 2^m$ and points x_n have been defined for each $1 \leq n < 2^{m-1}$ so that properties (a), (b) and (c) are valid for all n such that $1 \leq n < 2^{m-1}$. Suppose k satisfies $2^{m-1} \leq k < 2^m$. By hypothesis, we have diam $U_k > \epsilon$, hence there are functionals z_0^* and z_1^* in U_k such that $\|z_0^* - z_1^*\| > \epsilon$. Choose points $x_k \in E$ such that $\|x_k\| = 1$ and $\langle z_0^* - z_1^*, x_k \rangle = \epsilon + \delta$ for some $\delta > 0$. Let

$$U_{2k} = \{x^* \in U_k : \langle x^*, x_k \rangle > \langle z_0^*, x_k \rangle - \delta/2\}$$

$$U_{2k+1} = \{y^* \in U_k : \langle y^*, x_k \rangle < \langle z_1^*, x_k \rangle + \delta/2\}.$$

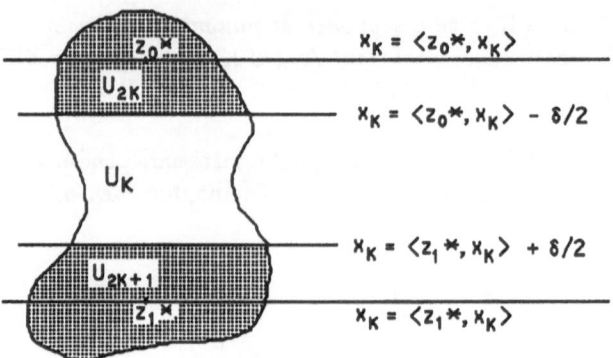

$$x_K = \langle z_0*, x_K \rangle$$

$$x_K = \langle z_0*, x_K \rangle - \delta/2$$

$$x_K = \langle z_1*, x_K \rangle + \delta/2$$

$$x_K = \langle z_1*, x_K \rangle$$

Fig. 5.2

Clearly, U_{2k} and U_{2k+1} are nonempty relatively weak* open subsets of A and U_{2k}, U_{2k+1} and x_k satisfy (a), (b) and (c) for $n = k$. Since $n = 2k$ and $n = 2k + 1$ exhaust $\{n : 2^{m-1} < n < 2^m\}$, the construction is complete.

We will shortly need the following elementary fact: If J_1, J_2 are nonempty subsets of a topological vector space, then

$$\overline{co}J_1 - \overline{co}J_2 \subset \overline{co}(J_1 - J_2),$$

where $\overline{co}$ denotes the closed convex hull. [If $a \in J_1$, then the fact that the map $x \to a - x$ is an affine homeomorphism shows that

$$a - \overline{co}J_2 = \overline{co}(a - J_2) \subset \overline{co}(J_1 - J_2).$$

For any $b \in \overline{co}J_2$, we also have $\overline{co}J_1 - b \subset \overline{co}(J_1 - J_2)$, whence the result.]
Now, for each n, let K_n denote the weak* closed convex hull of U_n. Each K_n is nonempty, weak* compact and convex, and (b) implies that $K_{2n} \cup K_{2n+1} \subset K_n$. By Lemma 5.5, there is a tree $\{x_n^*\}$ in E^* such that $x_n^* \in K_n$ for each n. By the previous remark, we necessarily have

$$x_{2n}^* - x_{2n+1}^* \in K_{2n} - K_{2n+1} \subset w^* - \text{cl } co(U_{2n} - U_{2n+1}),$$

so (c) implies that $\langle x_{2n}* - x_{2n+1}^*, x_n \rangle \geq \epsilon$. Hence, $\|x_{2n}^* - x_{2n+1}^*\| \geq \epsilon$ and $\{x_n^*\}$ is an infinite $\epsilon/2$-tree in $K_1 = w^*$-cl co$A$.

Theorem 5.7. *A Banach space E is an Asplund space if and only if E^* has the RNP.*

Proof. As we have observed, Theorem 2.31 contains the "only if" assertion. To prove the converse, suppose that E is not an Asplund space. By Theorem 2.33 there exists a separable subspace F of E such that F^* is not separable, hence Lemma 5.4 and Prop. 5.6 together show that E^* contains a bounded δ-tree, for some $\delta > 0$. Since the latter is not dentable, the space E^* does not have the RNP.

If C is a bounded closed convex subset of a Banach space with the RNP (or more generally, if C itself has the RNP) one can deduce a number of strong implications concerning the extremal structure of C; for instance, such a set is the closed convex hull of its extreme points, a property usually only associated (via the Krein-Milman theorem) with compact–or weakly compact–convex sets. More generally, we will eventually show that any such set is the closed convex hull of its *strongly exposed points*.

Definition 5.8. A point x in a closed convex set C is said to be *strongly exposed* provided there exists $x^* \neq 0$ such that $x \in S(x^*, C, \alpha)$ for every $\alpha > 0$ and these slices have diameters converging to 0 as α tends to 0. Equivalently, x is strongly exposed by x^* if, for $\{x_n\} \subset C$,

$$\langle x^*, x_n \rangle \to \sigma_C(x^*) \text{ implies } \|x_n - x\| \to 0.$$

A functional x^* which satisfies the definition above is called a *strongly exposing functional*, and is said to *strongly expose* x.

A point $x \in C$ is called *exposed* if there exists $x^* \neq 0$ such that

$$\langle x^*, x \rangle = \sigma_C(x^*) \text{ and } \langle x^*, y \rangle < \sigma_C(x^*) \text{ for each } y \in C, \quad y \neq x,$$

and x^* is said to *expose* x. If $C \subset E^*$, a point $x^* \in C$ is said to be *weak* strongly exposed* (or *weak* exposed*) if analogous properties hold, with the exposing functionals coming from E.

It is obvious that exposed points are extreme points and that strongly exposed points are exposed; the following simple examples show that both inclusions can be proper.

5.9. Examples.

Let C be the closed convex hull of the orthogonal basis vectors $\{e_n\}$ in ℓ^2. Then 0 is an exposed point of C, but every slice of C which contains 0 has diameter at least $\sqrt{2}$, so x is not strongly exposed.

Proof. Since $\{e_n\}$ converges weakly to 0 and C is weakly closed, we know that $0 \in C$. Note that if $y = (y_n) \in C$, then $y_n \geq 0$ for all n. This implies that if $x = (x_n) \in \ell^2$ is such that $x_n < 0$ for all n, then necessarily $\langle x, y \rangle < 0$ whenever $y \in C$, $y \neq 0$, which says that x exposes C at 0. On the other hand, any slice of C which contains 0 is a relative weak neighborhood of 0, hence contains all but finitely many basis vectors, and $\|e_n - e_m\| = \sqrt{2}$ whenever $m \neq n$.

To see that extreme points need not be exposed, consider the following sketch of a compact convex subset C of $\mathbf{R}^2$; the point of tangency x is an extreme point, but is not exposed.

Fig. 5.3

We next take a look at the duality relationship between Fréchet differentiability of the norm in E and weak* strongly exposed points of the dual ball in E^*; this was first investigated by V. L. Šmulyan in 1940. That such a duality is plausible is suggested by Fig. 5.3: If we assume that C is centered at the origin, it defines a unit ball in $\mathbb{R}^2$. The corresponding dual unit ball (that is, its polar) will have the form

Fig. 5.4

where x^* is the unique functional supporting C at x. Note that the flat (hence very smooth) portions of C are in duality with the pointed (hence very rotund) portions of its polar set.

A slightly more general formulation of Šmulyan's results can be obtained in terms of continuous nonnegative sublinear functionals (Minkowski gauges) and their "dual balls". Recall some basic facts from Sec. 1: A real-valued function p is *sublinear* provided $p(x+y) \leq p(x) + p(y)$ and $p(tx) = t \cdot p(x)$ for all $x, y \in E$ and $t \geq 0$. It is continuous if and only if there exists $M > 0$ such that $p(x) \leq M \cdot \|x\|$ for all x. A *nonnegative* continuous sublinear functional is called a *Minkowski functional* (or *gauge*). It can be characterized in terms of its associate closed convex "ball" $\{x \in E : p(x) \leq 1\}$, but we will be more interested in its "dual ball" $C(p)$ – or simply C – defined by

$$C(p) = \{x^* \in E : \langle x^*, x \rangle \leq p(x) \text{ for all } x \in E\}.$$

This is easily seen to be weak* compact and convex. The following elementary facts are useful enough that we spell out the details.

Lemma 5.10. *If p is a Minkowski gauge, then $x^* \in \partial p(x)$ if and only if $x^* \in C$ and $\langle x^*, x \rangle = p(x)$. Moreover, $p = \sigma_C$, the support functional for the set C.*

Proof. By definition, $x^* \in \partial p(x)$ provided

$$\langle x^*, y - x \rangle \leq p(y) - p(x) \text{ for all } y \in E. \tag{5.1}$$

Assuming this, apply it to the point ry where $r > 0$ and y is fixed, divide by r and let $r \to \infty$ to get $\langle x^*, y \rangle \leq p(y)$ for all y. Thus x^* is in C. Moreover, setting y in (5.1) equal to 0 yields $\langle x^*, x \rangle \geq p(x)$, so equality holds at x. Conversely, if $x^* \leq p$ and $\langle x^*, x \rangle = p(x)$, then (5.1) is immediate. It is also easy to see that

$$p(x) = \sigma_C(x) \equiv \sup\{\langle x^*, x \rangle : x^* \in C\};$$

indeed, if $x \in E$, then $\langle x^*, x \rangle \le p(x)$ for all $x^* \in C$, so $\sigma_C(x) \le p(x)$. Since p is continuous, there exists an element $x^* \in \partial p(x)$, so $x^* \in C$ and $p(x) = \langle x^*, x \rangle \le \sigma_C(x)$.

The duality property in which we are interested is described by the following proposition.

Proposition 5.11. *Suppose that p is a Minkowski gauge on E. An element x^* in $C(p)$ is weak* strongly exposed by $x \in E$ if and only if p is Fréchet differentiable at x, with $p'(x) = x^*$.*

Proof. Suppose that $x^* \in C$ is weak* strongly exposed by $x \ne 0$, so that $\langle x^*, x \rangle = \sigma_C(x) = p(x)$. Since p is continuous, $\partial p(y) \ne \emptyset$ for all y in E. Let $\varphi(y)$ be any element in $\partial p(y)$ if $y \ne x$, and let $\varphi(x) = x^*$. By Prop. 2.8, it suffices to show that φ is norm-to-norm continuous at x. Suppose, then, that $x_n \to x$. Since, for all n, $\varphi(x_n) \in \partial p(x_n)$, we have $\varphi(x_n) \in C$ and $\langle \varphi(x_n), x_n \rangle = p(x_n)$. Moreover,

$$|\langle \varphi(x_n) - x^*, x \rangle| = |\langle \varphi(x_n) - x^*, x - x_n \rangle + \langle \varphi(x_n), x_n \rangle - \langle x^*, x_n \rangle| \le$$

$$\le \|\varphi(x_n) - x^*\| \cdot \|x - x_n\| + |p(x_n) - \langle x^*, x_n \rangle|.$$

Since $p(x_n) \to p(x)$ and $\langle x^*, x_n \rangle \to \langle x^*, x \rangle = p(x)$, and $\|\varphi(x_n) - x^*\|$ is bounded, the right side converges to 0. Therefore

$$\langle \varphi(x_n), x \rangle \to \langle x^*, x \rangle = p(x).$$

Since x^* is weak* strongly exposed by x, we have $\|\varphi(x_n) - x^*\| \to 0$, so φ is norm-to-norm continuous at x; moreover, $x^* = p'(x)$.

To prove the converse, suppose that p is Fréchet differentiable at $x \ne 0$. Let $x^* = p'(x)$; since $x^* \in \partial p(x)$ we have $x^* \le p$ and $\langle x^*, x \rangle = p(x)$. Suppose that there exists $\{x_n^*\} \subset C$ and $r > 0$ such that

$$\langle x_n^*, x \rangle \to \langle x^*, x \rangle = p(x) \text{ but } \|x_n^* - x^*\| > r \text{ for all } n.$$

There exist $\{y_n\} \subset E$ such that $\|y_n\| = 1$ and

$$\langle x_n^* - x^*, y_n \rangle > r \text{ for all } n. \tag{5.2}$$

By Fréchet differentiability, given $0 < \epsilon < r/2$ there exists $\delta > 0$ such that

$$0 \le p(x + y) - p(x) - \langle x^*, y \rangle \le \epsilon \|y\|$$

whenever $\|y\| \le \delta$. Since $x_n^* \in C$ we have $\langle x_n^*, x + y \rangle \le p(x + y)$ for all $y \in E$. Let $z_n = \delta y_n$; then from (5.2) we have

$$\delta r < \langle x_n^* - x^*, z_n \rangle = \langle x_n^*, x + z_n \rangle - \langle x^*, x \rangle - \langle x^*, z_n \rangle - \langle x_n^* - x^*, x \rangle \le$$

$$\le [p(x + z_n) - p(x) - \langle x^*, z_n \rangle] - \langle x_n^* - x^*, x \rangle \le \delta r/2 - \langle x_n^* - x^*, x \rangle,$$

which leads to a contradiction, since $\langle x_n^* - x^*, x \rangle \to 0$.

This proposition is the basis for the following dual characterization of Asplund spaces.

Theorem 5.12. *A Banach space E is an Asplund space if and only if every nonempty weak* compact convex subset C of E^* is the weak* closed convex hull of its weak* strongly exposed points.*

Proof. One direction is immediate from Theorem 2.32: If every such subset $C \subset E^*$ has weak* strongly exposed points, then every bounded nonempty subset A of E^* is weak* dentable (since small weak* slices of the weak* closed convex hull C of A yield small weak* slices of A). To prove the converse, suppose that E is an Asplund space and that C is a nonempty weak* compact convex subset of E^*. Without loss of generality, $0 \in C$. Let $p = \sigma_C$; then p is a Minkowski gauge and – as the definition and an easy application of the separation theorem show – $C(p) = C$. Let D be the weak* closed convex hull of the weak* strongly exposed points of C and suppose that $D \neq C$. Then there exists $x \in E$ such that $\sigma_D(x) < \sigma_C(x)$. Since C is bounded, both of these convex support functionals are continuous and so there exists a point of Fréchet differentiability x_0 of p contained in the open set where the strict inequality holds: $\sigma_D(x_0) < \sigma_C(x_0)$. By Proposition 5.11, x_0 strongly exposes C at the point $x^* = p'(x_0)$ and so $\langle x^*, x_0 \rangle = p(x_0) = \sigma_C(x_0)$. But D contains all the weak* strongly exposed points of C, therefore $\langle x^*, x_0 \rangle \leq \sigma_D(x_0) < \sigma_C(x_0)$, a contradiction.

As another application of Proposition 5.11, we can now give the details of Example 1.14(c).

Example 1.14(c).

There exists an equivalent norm on ℓ^1 which is nowhere Fréchet differentiable but which is Gâteaux differentiable at each nonzero point.

Proof. By Exercise 2.37(a), to obtain an equivalent smooth norm on ℓ^1 it suffices to define an equivalent strictly convex dual norm on ℓ^∞. To that end, define $\| \cdot \|$ on ℓ^∞ by

$$\|y\| = \|y\|_\infty + p(y), \text{ where } p(y) = (\Sigma 2^{-n} y_n^2)^{1/2}.$$

Clearly, $\|y\|_\infty \leq \|y\| \leq 2\|y\|_\infty$ for each $y \in \ell^\infty$, so this is an equivalent norm. Since p is the supremum of a sequence of weak* continuous functions, it is weak* lower semicontinuous, so $\| \cdot \|$ is a *dual* norm. As the sum of two norms, one of which (namely, p) is strictly convex, $\| \cdot \|$ is easily verified to be strictly convex. It remains to show that the norm $\| \cdot \|$ which it induces on ℓ^1 is nowhere Fréchet differentiable, and this will follow from Prop. 5.12 if we show that the unit ball it defines in ℓ^∞ has no weak* strongly exposed points. Suppose, then, that $x \in \ell^1$ with $\|x\| = 1$ and that $y \in \ell^\infty$, $\|y\| = 1$, is the unique element such that $\langle y, x \rangle = 1$. It suffices to produce a sequence $\{y^k\}$ in ℓ^∞ such that

$$\langle y^k, x \rangle \to \langle y, x \rangle = 1 \text{ and } \|y^k\| \to \|y\| = 1$$

but $\|y^k - y\|$ does not converge to zero. There are two cases to consider, depending on whether $y_n \to 0$. If the latter *is* the case, then there exists N such that $|y_n| < 1/4$ whenever $n > N$. For $k > N$, define $y^k \in \ell^\infty$ by

$$(y^k)_n = y_n \text{ if } n \neq k, (y^k)_k = 3/8.$$

Since $x_k \to 0$, we have

$$\langle y^k, x \rangle = \langle y, x \rangle - x_k \cdot (y_k - 3/8) \to \langle y, x \rangle = 1.$$

Also, $1 = \|y\| \leq 2\|y\|_\infty$ so $\|y\|_\infty \geq 1/2$. Thus,

$$\|y^k\|_\infty = \max\{3/8, \sup_{n \neq k} |y_n|\} = \|y\|_\infty.$$

It follows that for $k \to \infty$,

$$\|y^k\| = \|y\|_\infty + [\Sigma 2^{-n} y_n^2 - (y_k^2 - 9/64) \cdot 2^{-k}]^{1/2} \to \|y\|.$$

On the other hand, $\|y^k - y\| \geq \|y^k - y\|_\infty = |3/8 - y_k| > 1/8$. Suppose, now, that $\{y_n\}$ does *not* converge to zero. There necessarily exist $\epsilon > 0$ and a subsequence $\{y_{n_k}\}$ such that $|y_{n_k}| > 2\epsilon$ for all k. Define y^k in ℓ^∞ by

$$(y^k)_n = y_n \text{ if } n \neq n_k, (y^k)_{n_k} = \epsilon.$$

Then $\langle y^k, x \rangle \to \langle y, x \rangle = 1$ and $\|y^k\| \to \|y\|$ as before, while

$$\|y - y^k\| \geq \|y - y^k\|_\infty = |y_{n_k} - \epsilon| > \epsilon,$$

which completes the proof.

Some of the results in this and earlier sections can be interpreted as theorems in infinite dimensional optimization, in that they give existence results for minimum (or maximum) points for certain types of functions. Actually, what they show is better described as "perturbed" optimization. In finite dimensional optimization, existence is frequently trivial: When one has a continuous (or lower-bounded lower semicontinuous) real-valued function f on a bounded and closed set C, then, since the latter is compact, the existence of a minimum point is guaranteed. In the infinite dimensional case, minimum points need not exist, and one will settle for a theorem of the form: For some Lipschitz function h, having Lipschitz norm as small as one may wish, the perturbed function $f + h$ attains its minimum on C. Ekeland's Lemma 3.13 can be cast in this form (as noted just preceding its statement). So can the Brøndsted-Rockafellar Theorem 3.17, which has the following consequence: Suppose that f is a convex proper lower semicontinuous function which nearly attains its minimum on the Banach space E at the point x. Then there is a continuous linear functional x^* of small norm such that $f + x^*$ does attain its minimum. (Moreover, it does so at a point which is close to x.) In these results, compactness was replaced by completeness and convexity. Results of this character have been obtained by C. Stegall [Ste$_2$] (who replaces compactness by convexity and the Radon-Nikodým property) and by N. Ghoussoub and B. Maurey [Gh-M$_2$] (who look at a certain class of subsets of dual Banach spaces and replace compactness by weak* metrizability and weak* relative compactness). We will formulate and prove Stegall's theorem in terms of maximum points for upper-bounded and upper semicontinuous functions. (The theorems of Ghoussoub and Maurey will be stated without proof at the end of this section.) Our proof of Stegall's theorem requires a straightforward extension of the notion of "slice", so as to include certain nonlinear functions.

Definition 5.13. Suppose that C is a nonempty set and that f is an upper-bounded real-valued function on C. For each $\alpha > 0$ define

$$S(f, C, \alpha) = \{x \in C : f(x) > \sup_C f - \alpha\}.$$

A point $x \in C$ is said to be a *strong maximum* for f if $f(x) = \sup_C f$ and $\|x - x_n\| \to 0$ whenever $f(x_n) \to f(x)$. (Note that we could have said that f "strongly exposes" x; some authors use this terminology.)

5.14. Exercise.

If C is closed and nonempty and f is a real-valued upper-bounded upper semi-continuous function on C, then f attains a strong maximum on C if and only if diam $S(f, C, \alpha) \to 0$ as $\alpha \to 0^+$.

Theorem 5.15. (Stegall). *Suppose that $C \subset E$ is a nonempty closed and bounded convex set with the RNP and that f is an upper-bounded upper semi-continuous real-valued function on C. Given $\epsilon > 0$, there exists x^* in E^* such that $\|x^*\| \leq \epsilon$ and $f + x^*$ attains a strong maximum on C.*

Remark. An obvious difference between this theorem and Ekeland's variational principle is that while both have the same hypotheses on the function to be minimized (or maximized), Stegall's result obtains a maximum point after adding a small *linear* function (which is also Lipschitz, of course). This strengthening of Ekeland's variational principle comes at the cost of assuming that C has the RNP.

Before we start the proof of Theorem 5.15 we need to recall the original definition of dentability, in the form of the following exercise (which is a simple application of the separation theorem). We will also require several lemmas.

5.16. Exercise.

Prove that a nonempty closed set A in a Banach space E is dentable if and only if for every $\epsilon > 0$ there exists $x \in A$ such that x is not in $\overline{co}(A \setminus B(x; \epsilon))$.

The next lemma gives a recipe for constructing nondentable sets; we will use it when we prove Theorem 5.15 by contradiction.

Lemma 5.17. *Suppose that $\{A_n\}$ is a sequence of (eventually) nonempty subsets of a Banach space E with the following property: There exist constants $\epsilon > 0$ and $\lambda > 0$ such that for all $x \in co\, A_n$ and $y \in E$,*

$$dist\,[x,\ co(A_{n+1} \setminus B(y; \epsilon))] \le \lambda/2^n.$$

Then the set

$$A = \cap_n \overline{\cup\{co\,A_j : j \ge n\}}$$

is nonempty and not dentable.

Proof. We will show that $x \in \overline{co}(A \setminus B(x; \epsilon/2))$ for each $x \in A$. First however, we show that A is nonempty and that for each $n \ge 1$,

$$co\,A_n \subset A + (4\lambda/2^n)B, \tag{5.3}$$

where $B \equiv B(0; 1)$. To this end, fix n sufficiently large that A_n is nonempty and suppose that $x_0 \in co\,A_n$. By hypothesis, there exists a point $x_1 \in co\,A_{n+1}$ such that $\|x_0 - x_1\| \le 2\lambda/2^n$. Similarly, there exists a point $x_2 \in co\,A_{n+2}$ such that $\|x_1 - x_2\| \le 2\lambda/2^{n+1}$. Continuing by induction, we obtain $x_k \in co\,A_{n+k}$, $k = 0, 1, 2, \ldots$ such that $\Sigma\|x_k - x_{k+1}\| < \infty$. This implies that the series $\Sigma(x_k - x_{k+1})$ converges to an element $y \in E$, of norm at most $4\lambda/2^n$. By writing y as a limit of the (collapsing) partial sums we get $y = x_0 - z$ where $z = \lim x_k$. It follows that $z \in A$ (so A is nonempty) and $x_0 \in A + (4\lambda/2^n)B$, which proves (5.3). Suppose that $x \in A$. Fix m such that $4\lambda/2^m < \epsilon/2$. Since we have $x \in \{\cup co\,A_j : j \ge n\}$ for *all* n, for each $n \ge m$ there exist $j \ge n$ and $y_n \in co\,A_j$ such that $\|x - y_n\| \le \lambda/2^n$. By hypothesis,

$$dist\,[y_n,\ co(A_{j+1} \setminus B(x; \epsilon))] \le \lambda/2^j \le \lambda/2^n,$$

so there exists $z_n \in co(A_{j+1} \setminus B(x; \epsilon))$ such that $\|y_n - z_n\| < 2\lambda/2^n$. We can write z_n as a finite convex combination

$$z_n = \Sigma\lambda_i u_i, \qquad u_i \in A_{j+1} \setminus B(x; \epsilon).$$

By (5.3), for each i we have $u_i \in A + (4\lambda/2^j)B \subset A + (4\lambda/2^n)B$, so there exists $v_i \in A$ such that $\|u_i - v_i\| \le 4\lambda/2^n$. Let $w_n = \Sigma\lambda_i v_i$; it follows that $\|z_n - w_n\| \le 4\lambda/2^n$ and hence $\|x - w_n\| \le 7\lambda/2^n$. Since $\|u_i - x\| \ge \epsilon$ for each i, we have

$$\|v_i - x\| \ge \epsilon - 4\lambda/2^n \ge \epsilon - 4\lambda/2^m > \epsilon/2,$$

that is, $w_n \in co(A \setminus B(x; \epsilon/2))$. It follows that $x \in \overline{co}(A \setminus B(x; \epsilon/2))$ as required.

The next two lemmas will allow us to reduce the proof of Theorem 5.13 to simply showing that there are arbitrarily small perturbations $f + x^*$ of f which define small slices. The first lemma shows that we can get decreasing families of slices provided we perturb the function f slightly; the proof is a straightforward verification from the definition. Since all our slices will be in the same set C, we will write $S(f, \alpha)$ in place of the usual $S(f, C, \alpha)$.

Lemma 5.18. *Suppose that the real-valued function f is bounded above on the nonempty subset C of the unit ball B of E. For any $\alpha > 0$ we have $S(f + x^*, \beta) \subset S(f, \alpha)$ provided $\|x^*\| < \alpha/2$ and $0 < \beta < \alpha - 2\|x^*\|$.*

Lemma 5.19. *Let C be a nonempty bounded closed subset of E and suppose that for every upper-bounded upper semicontinuous function f on C and every $\epsilon > 0$, there exist $x^* \in E^*$, $\|x^*\| < \epsilon$, and $\alpha > 0$ such that diam $S(f + x^*, \alpha) \leq 2\epsilon$. Then given $\epsilon > 0$ there exists $x^* \in E^*$ such that $\|x^*\| < \epsilon$ and $f + x^*$ attains a strong maximum on C.*

Proof. We assume without loss of generality that C is contained in the unit ball of E and that $0 < \epsilon < 1$. By hypothesis, there exist $x_1^* \in E^*$, $\|x_1^*\| < \epsilon/2$, and $0 < \alpha_1 < 1$ such that diam $S_1 < \epsilon$, where we write $S_1 = S(f + x_1^*, \alpha_1)$. By applying the hypotheses to $f + x_1^*$ and $\epsilon_1 = \alpha_1 \cdot \epsilon/2^2$, we obtain $\|x_2^*\| < \epsilon_1$ so that diam $S_2 < 2\epsilon_1$, where $S_2 = S(f + x_1^* + x_2^*, \alpha_2)$ and $0 < \alpha_2 < \alpha_1$. Continuing by induction with $\alpha_0 = 1$, we obtain sequences

$$\epsilon_n > 0, \quad 0 < \alpha_n < 1, \quad x_n^* \in E^*, \quad S_n = S(f + \Sigma_1^n x_k^*, \ \alpha_n)$$

such that

$$\text{diam } S_n \leq 2\epsilon_{n-1}, \quad \|x_n^*\| < \epsilon_{n-1}, \quad \epsilon_n = \epsilon \cdot \alpha_n/2^{n+1} \text{ and } \alpha_n < \alpha_{n-1}.$$

Obviously, the series Σx_n^* converges to a point $x^* \in E^*$ of norm at most $\epsilon \cdot \Sigma \alpha_n 2^{-(n+1)} < \epsilon$. Note that diam $S_n \to 0$, since $\epsilon_n \to 0$. We claim $f + x^*$ attains a strong maximum on C. By Exercise 5.14 it suffices to show that diam $S(f + x^*, \alpha) \to 0$ as $\alpha \to 0^+$, and for this it suffices to prove that for all n there exists $\alpha > 0$ such that $S(f + x^*, \alpha) \subset S_n$. This follows from Lemma 5.18, using the fact that $f + x^* = f + \Sigma_1^n x_k^* + w_n^*$ where

$$\|w_n^*\| \leq \Sigma_{k=n+1}^\infty \epsilon \cdot \alpha_{k-1}/2^k < \alpha_n/2;$$

we need only choose α such that $0 < \alpha < \alpha_n - 2\|w_n^*\|$.

Proof of Theorem 5.15. By utilizing the previous lemma, we need only show that given any $\epsilon > 0$ there exist $x^* \in E^*$, $\|x^*\| < \epsilon$, and $\alpha > 0$ such that diam $S(f + x^*, \alpha) \leq 2\epsilon$. Proceeding by contradiction, suppose that for every $\|x^*\| < \epsilon$ and each $\alpha > 0$, we have diam $S(f + x^*, \alpha) > 2\epsilon$. For each n let

$$A_n = \cup\{S(f + x^*, 1/4^n): \|x^*\| \leq \epsilon - 2^{-n}\}.$$

For all sufficiently large n, we have $\epsilon - 2^{-n} > 0$, so that A_n is nonempty. Let $\lambda = 5/2$; we will show that with this choice of λ, the sequence $\{A_n\}$ satisfies the hypotheses of Lemma 5.17, the conclusion of which will contradict the hypothesis of Theorem 5.15 (that C has the RNP). Restating the main hypothesis of Lemma 5.17, we want to show that for any $y \in E$,

$$\text{co } A_n \subset \text{co}(A_{n+1} \setminus B(y; \epsilon)) + (\lambda/2^n) \cdot B. \tag{5.4}$$

Since the set on the right side is convex, it suffices to prove that it contains A_n. Suppose, then, that $x \in A_n$ but, for some $y \in E$, it is not in the right hand side of (5.4) (which has nonempty interior). By the separation theorem there exists $y^* \in E^*$, $\|y^*\| = 1$, such that

$$\langle y^*, x \rangle \geq \sup\{\langle y^*, u \rangle : u \in A_{n+1} \setminus B(y; \epsilon)\} + \lambda/2^n. \tag{5.5}$$

Since $x \in A_n$, there exists $x^* \in E^*$ with $\|x^*\| \leq \epsilon - 2^{-n}$ such that x is in the slice $S(f + x^*, 1/4^n)$. Write

$$z* = x^* + (1/2^{n+1}) \cdot y^*;$$

it follows that $\|z^*\| \leq \epsilon - 2^{-n} + 2^{-(n+1)} = \epsilon - 2^{-(n+1)}$ so $S(f + z^*, 1/4^{n+1}) \subset A_{n+1}$. Since diam $S(f+z^*, 1/4^{n+1}) > 2\epsilon$, this slice is not contained in $B(y; \epsilon)$. This implies that there exists $z \in C \setminus B(y; \epsilon)$ such that

$$f(z) + \langle z^*, z \rangle > \sup_C (f + z^*) - 1/4^{n+1}. \tag{5.6}$$

Thus, $z \in A_{n+1} \setminus B(y; \epsilon)$ and hence, from (5.5), we conclude that

$$\langle y^*, x \rangle \geq \langle y^*, z \rangle + \lambda/2^n. \tag{5.7}$$

We will prove the inclusion in (5.4) by showing that this cannot be true. Note first that since $x \in S(f + x^*, 1/4^n)$ and $z \in C$, we necessarily have

$$f(x) + \langle x^*, x \rangle > \sup_C (f + x^*) - 1/4^n \geq f(z) + \langle x^*, z \rangle - 1/4^n. \tag{5.8}$$

Similarly, from (5.6) we have

$$f(z) + \langle z^*, z \rangle \equiv f(z) + \langle x^*, z \rangle + (1/2^{n+1})\langle y^*, z \rangle >$$
$$> \sup_C[f + x^* + (1/2^{n+1})y^*] - 1/4n + 1 \geq \tag{5.9}$$
$$\geq f(x) + \langle x^*, x \rangle + (1/2^{n+1})\langle y^*, x \rangle - 1/4^{n+1}.$$

Using (5.9) and (5.8) we obtain

$$f(z) + \langle x^*, z \rangle + (1/2^{n+1})\langle y^*, z \rangle >$$
$$> f(z) + \langle x^*, z \rangle - 1/4^n + (1/2^{n+1})\langle y^*, x \rangle - 1/4n + 1$$

or

$$(1/2^{n+1})\langle y^*, x - z \rangle < 1/4^n + 1/4^{n+1} = 5/4^{n+1} = 5/2^{2n+2}.$$

Equivalently, $\langle y^*, x - z \rangle < 5/2^{n+1} = (5/2)(1/2^n) = \lambda/2^n$, which contradicts (5.7) and completes the proof.

We next illustrate the power of Theorem 5.15 by using it to give a simple proof of the geometric fact mentioned earlier, that RNP sets are generated by their strongly exposed points.

Theorem 5.20. *Suppose that $C \subset E$ is a nonempty bounded closed convex set with the RNP. Then C is the closed convex hull of its strongly exposed points. Moreover, the functionals which strongly expose points of C constitute a dense G_δ subset of E^*.*

Proof. We first prove the second assertion. To this end, define

$$G_n = \{x^* \in E^* \colon \text{diam } S(x^*, C, \alpha) < 1/n \text{ for some } \alpha > 0\}.$$

If $y^* \in G_n$ (so that diam $S(x^*, C, \alpha) < 1/n$ for some α), then by Lemma 5.18, applied to $f = y^*$, the $\alpha/2$-neighborhood of y^* is contained in G_n, hence the latter is open. Theorem 5.15 (applied to any element f of E^*) trivially implies that each G_n is dense in E^*, hence by the Baire category theorem, the set $G = \cap G_n$ is a dense G_δ subset of E^*. Obviously, if x^* strongly exposes C, then it defines slices of arbitrarily small diameter, hence is in G. Conversely, if $x^* \in G$, then it will define a nested sequence $\{S(x^*, C, \alpha_n)\}$ of slices of C with diameters converging to 0. Their closures intersect in a point of C which is strongly exposed by x^*. To prove the first assertion of the theorem, let D be the closed convex hull of the strongly exposed points of C. If $D \neq C$, then by the separation theorem there exists $x^* \epsilon\ E^*$ such that $\sigma_D(x^*) < \sigma_C(x^*)$. Since these functionals are norm continuous on E^*, there exists a functional in the dense set G for which the same inequality holds, contradicting the definition of D.

Since strongly exposing functionals can be used to define slices of arbitrarily small diameter, the foregoing theorem leads to a nice characterization.

Theorem 5.21. *A Banach space E has the RNP if and only if every bounded closed convex subset of E is the closed convex hull of its strongly exposed points.*

The following variant (and corollary) of Theorem 5.15 due to M. Fabian [Fa$_2$] yields a strong minimum for a lower semicontinuous function f defined on a space with the RNP. It replaces the restriction that f be defined only on a bounded set with a strong lower-boundedness hypothesis for f on all of E.

Corollary 5.22. (Fabian). *Suppose that the Banach space E has the RNP and that $f \colon E \to \mathbf{R} \cup \{\infty\}$ is a lower semicontinuous function on E for which there exist $a > 0$ and $b \in \mathbf{R}$ such that*

$$f(x) > 2a\|x\| + b, \quad x \in E.$$

Then for any $\epsilon > 0$ there exists $x^* \in E^*$ such that $\|x^*\| < \epsilon$ and $f + x^*$ attains a strong minimum on E.

Proof. Since we can replace f by $f - b$, we may assume that $b = 0$. Note that if $x^* \in E^*$ with $\|x^*\| < a$, then for any $x \in E$

$$f(x) + \langle x^*, x \rangle \geq 2a\|x\| - a\|x\| = a\|x\|. \tag{5.10}$$

Let $r = (1/a)[f(0) + 1]$ and apply Theorem 5.15 to $-f$ restricted to the ball $B \equiv B(0; r)$, with $0 < \epsilon < a$. Thus, there exist $x^* \in E^*$, $\|x^*\| < \epsilon$, and a point $x_0 \in B$ such that $f + x^*$ attains a strong minimum in B at x_0. We need only show that this is a strong minimum for $f + x^*$ in E. If $x \in E$ is such that $f(x) + \langle x^*, x \rangle \leq f(x_0) + \langle x^*, x_0 \rangle = \inf_B(f + x^*) \leq f(0)$, then from (5.10) we conclude that $\|x\| \leq (1/a)f(0) < r$, so $x = x_0$. Similarly, if $\{x_n\} \subset E$ and

$(f + x^*)(x_n) \to (f + x^*)(x_0)$, then eventually $f(x_n) + \langle x^*, x_n \rangle < f(0) + 1$, so from (5.10) it follows that $x_n \in B$ and therefore $x_n \to x_0$.

We now describe without proof some of the work of Ghoussoub and Maurey on perturbed optimization; it involves the following notion.

Definition 5.23. A bounded nonempty subset $C \subset E^*$ is said to be a *strong w^*-H_δ set* if its complement in its weak* closed convex hull D is the union of a sequence $\{K_n\}$ of weak* compact convex sets, each of which is of positive distance from C.

The reader is referred to [Gh-M₂] for the motivation for this definition; we mention only that weak* compact convex subsets of E^* obviously have this property, and many of the extremal properties of such sets are shown to carry over to strong w^*-H_δ sets (and related families of sets).

Theorem 5.24. *Suppose that the nonempty bounded set $C \subset E^*$ is a strong $w^* - H_\delta$ set in its weak* closed convex hull D and that D is weak* metrizable [or that C is norm separable]; then for any lower-bounded weak* lower semicontinuous function f on C and any $\epsilon > 0$, there exists $x \in E$ with $\|x\| < \epsilon$ such that $f + x$ attains a strict [strong] minimum on C.*

Note that in this result, the function f is perturbed by a *weak** continuous linear functional (which is necessarily Lipschitz).

Ghoussoub and Maurey also obtain a similar result for subsets of a Banach space E which has the following property.

Definition 5.25. A Banach space is said to have the *point of norm to weak continuity property* (PCP) if every nonempty bounded subset of E admits relative weak neighborhoods of arbitrarily small norm diameter. [Equivalently, every nonempty bounded closed subset contains at least one point at which the identity map (restricted to the subset) is norm to weak continuous; this explains the terminology.]

For more about the PCP, see [Ed-W] and [Gh-M₁]. It is clear that the RNP implies the PCP (slices define weak neighborhoods), so the following result should be compared to Theorem 5.15. In one sense, it is more general, since it applies to nonconvex sets; on the other hand, the perturbation R need not be linear nor do we get a strong minimum; also, E is assumed to be separable.

Theorem 5.26. *Suppose that the separable Banach space E has the PCP and that C is a closed bounded nonempty subset of E. If f is a lower-bounded lower semicontinuous function on C and $\epsilon > 0$, then there exists a norm Lipschitz and weakly continuous function h, of Lipschitz constant at most ϵ, such that $f + h$ attains its minimum on C.*

Finally, N. Ghoussoub, J. Lindenstrauss and B. Maurey [G-L-M] have shown that a *complex* Banach space E has the "analytic" RNP if and only if for every bounded upper semicontinuous real-valued function f on a closed bounded subset A of E and every $\epsilon > 0$, there is a *plurisubharmonic* function g on E, with supremum norm on A at most ϵ, such that $f + g$ attains a strong maximum on A. We refer to [G-L-M] for details and relevant definitions.

Remarks.

The fact that bounded subsets in the dual of an Asplund space admit weak* slices of arbitrarily small diameter was proved in [Na-Ph]. The converse (Theorem 5.7) was proved by C. Stegall [Ste₁]; the simpler proof given here is due to van Dulst and Namioka [Du-N]. The fact that convex sets with the RNP are generated by their strongly exposed points (Theorem 5.20) has been of considerable interest. It was first proved by the author [Ph₂], using geometrical methods and assuming that the entire space had the RNP. J. Bourgain then proved the general case [Bo], also using geometrical methods. C. A. Rogers pointed out that the original proof could easily be modified to yield the general result [La-Ph, p. 119]. K. Kunen and H. P. Rosenthal [Ku-R] have proved it using vector-valued martingales, not as outlandish as it appears, if one is aware that the RNP can be characterized in terms of a martingale convergence theorem (see [Bou]). A self-contained proof, using a Kenderov-like generic continuity theorem and the duality between differentiability and strongly exposed points, is given in [Ph₃]. We have presented the result here, of course, as a rather easy corollary to Theorem 5.15. Our proof of the latter is due to J. Bourgain [Bo₂], who used a modification of his proof that RNP sets can be characterized in terms of the so-called Bishop-Phelps property [Bo₁].

It is interesting to note that Theorem 5.15 is an easy consequence of Corollary 5.22, at least in the special case when the entire space E is assumed to have the RNP. (This is the case applied, for instance, in [Cr-Li₁,₂]). Indeed, if f is lower semicontinuous and lower bounded (by m say) on the bounded closed convex set C and if the latter is contained in a ball of radius $r > 0$, then the function which equals f in C and $+\infty$ outside C satisfies the hypotheses of Corollary 5.22, taking $a = 1/2$ and $b = m - r$. Fabian's proof in [Fa₂] of Corollary 5.22, apparently obtained independently of Stegall's (earlier) paper [Ste₂], uses the version of Theorem 5.20 found in [Ph₂], applied to certain subsets of $E \times \mathbf{R}$, and is much easier than the proof of Theorem 5.15 given here.

It follows trivially from Theorem 5.20 that a Banach space with the RNP has the *Krein-Milman Property* (KMP): Every bounded closed convex subset of E is the closed convex hull of its extreme points. *It remains an open question whether a Banach space with the KMP has the RNP.* There have been a number of partial results; for instance, R. Huff and P. D. Morris (see [Bou, p. 91]) have shown that the answer is affirmative in any dual space, J. Bourgain and M. Talagrand (also see [Bou, p. 423]) have shown the same for any Banach lattice while V. Caselles [Ca] has given a short proof of this and of C.–H Chu's result [Chu] that it is true in the predual of a von Neumann algebra. W. Schachermayer [Sch₁] has given an affirmative answer for Banach spaces which are isomorphic to their squares, as well as for convex sets which are "strongly regular" [Sch₂].

The reader has undoubtedly wondered whether the duality between Asplund spaces and the RNP goes the other way. It almost does: *A Banach space E has the RNP if and only if every continuous convex function on E^* which is also weak* lower semicontinuous is Fréchet differentiable at the points of a dense G_δ subset of E^*.* This was proved by J. Collier [Co] and generalized by S. Fitzpatrick [Fi]; see [Bou] for an exposition.

Section 6

6. Gâteaux differentiability spaces.

The following class of spaces is formally larger than the class of weak Asplund spaces, but in some ways is a more natural object of study.

Definition 6.1. A Banach space E is called a *Gâteaux Differentiability Space* (GDS) provided the set G of points of Gâteaux differentiability of a convex continuous function defined on a nonempty open convex subset $D \subset E$ is necessarily *dense* in D.

Clearly, a GDS differs from a weak Asplund space only by virtue of the fact that one does not require the set G to contain a dense G_δ. As shown in Prop. 1.25, the G_δ property is automatic for Fréchet differentiability, but known examples (see e.g., [Č-K] or [Tal₁]) show that for Gâteaux differentiability (of individual convex continuous functions, at least), it is definitely an additional requirement. In this section we will examine some properties of Gâteaux differentiability spaces, all but one of which remain open questions for weak Asplund spaces.

Another motivation for studying Gâteaux differentiability spaces is that they admit an interesting characterization which is completely analogous to Theorem 5.12 (which characterized Asplund spaces in terms of weak* strongly exposed points in their duals). We state this result now; its proof will be given below.

Theorem 6.2. *A Banach space E is a GDS if and only if every weak* compact convex subset of E^* is the weak* closed convex hull of its weak* exposed points.*

There is only one permanence (or stability) property known to be valid for weak Asplund spaces (they are preserved under quotient maps (Theorem 4.24)), while the Gâteaux differentiability spaces have that property (even a stronger property; see Prop. 6.8 below) plus a simple but useful stability property under certain products (which is still open for weak Asplund spaces).

It will help the exposition to introduce temporarily a class of spaces which is formally larger than the class of Gâteaux differentiability spaces (but which will turn out to be the same).

Definition 6.3. A Banach space E is said to be an *M-differentiability space* (MDS) provided every Minkowski gauge on E is Gâteaux differentiable at the points of a dense set.

Proposition 6.4. *If $E \times \mathbf{R}$ is an MDS, then the Banach space E is a GDS.*

Proof. We pair $E \times \mathbf{R}$ with its dual $E^* \times \mathbf{R}$ by

$$\langle (x^*, r^*), (x, r) \rangle = \langle x^*, x \rangle + r^* \cdot r.$$

Suppose, now, that f is a continuous convex function defined on an open convex subset D of E. We can assume without loss of generality that the origin is in D and that $f(0) = -1$. Let μ be the Minkowski functional defined on $E \times \mathbf{R}$ by the convex set epi(f); since the latter has nonempty interior [containing, for instance, the origin], μ is necessarily continuous. If μ is Gâteaux differentiable on a dense set, then by positive homogeneity it is necessarily Gâteaux differentiable on a dense subset of

$$G = \{(x, r) : x \in D \text{ and } \mu(x, r) = 1\}.$$

The set G is that part of the boundary of epi(f) whose "x-coordinate" is in D, so it is the graph of f and hence it is homeomorphic with D. Thus, it suffices to show that if μ is Gâteaux differentiable at the point $(x, f(x))$ (where $x \in D$), then f is Gâteaux differentiable at x. It is clear from Figure 6.1 below that any corners in the graph of f will be points at which epi(f) will have distinct supporting hyperplanes, hence points at which μ fails to be Gâteaux differentiable.

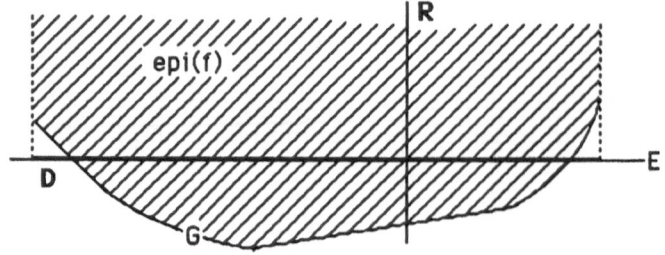

Fig. 6.1

Specifically, suppose that x_1^*, x_2^* were distinct subdifferentials to f at x, so that

$$\langle x_i^*, y - x \rangle \leq f(y) - f(x) \text{ for all } y \in D, \quad i = 1, 2. \tag{6.1}$$

Taking $y = 0$, we see that $\langle x_i^*, x \rangle - f(x) \geq -f(0) = 1$, so the reciprocal s_i of the quantity on the left is positive, for $i = 1, 2$. We claim that the two functionals $s_i(x_i^*, -1)$, $i = 1, 2$, are distinct subdifferentials to μ at $(x, f(x))$. It suffices to show that for each i,

$$\langle s_i(x_i^*, -1), (z, r) \rangle \leq \mu(z, r) \text{ for } (z, r) \in E \times \mathbf{R}, \tag{6.2}$$

with equality at $(z, r) = (x, f(x))$. By positive homogeneity of both sides of (6.2), it will be proved if we show that

$$\langle s_i(x_i^*, -1), (z, r) \rangle < 1$$

whenever $\mu(z, r) < 1$. The latter, of course, implies that (z, r) is interior to epi(f), so that $z \in D$ and $r > f(z)$. Thus, by using this fact and (6.1) with $y = z$,

$$s_i(\langle x_i^*, z \rangle - r) < s_i(\langle x_i^*, z \rangle - f(z)) \le s_i(\langle x_i^*, x \rangle - f(x)) = 1$$

as was to be shown. The equality of (6.2) at $(x, f(x))$ is just the definition of s_i, since $\mu(x, f(x)) = 1$. Finally, the two functionals $s_i(x_i^*, -1)$ are distinct: if $s_1(x_1^*, -1) = s_2(x_2^*, -1)$, then $s_1 = s_2$ and hence $x_1^* = x_2^*$.

Our stability result is the following:

Proposition 6.5. *If E is a GDS, then $E \times \mathbf{R}$ is a GDS.*

Proof. Suppose that D is an open convex subset of $E \times \mathbf{R}$, that f is a continuous convex function on D and that $(x_0, t_0) \in D$. Let U be an open neighborhood of (x_0, t_0); we want to find a point of Gâteaux differentiability of f in U. Without loss of generality we can suppose that U contains a neighborhood of the form $B \times I$, where B is an open ball centered at x_0 and $I = [t_0 - \delta, t_0 + \delta]$; we can also assume that $|f|$ is bounded in U, by M, say. Choose a differentiable extended real-valued function $g \le 0$ on I which equals $-\infty$ at the endpoints, is finite elsewhere and satisfies $g(t_0) = 0$. Define

$$h(x) = \sup\{f(x, t) + g(t) : t \in \mathbf{R}\}, \qquad x \in B.$$

Since h is the supremum of a family of convex functions, it is convex; moreover, since f is bounded above by M on B, so is h. Also, for any point $x \in B$, $h(x) \ge f(x, t_0) + g(t_0) \ge -M$, so h, being bounded, is continuous on B. (Recall the remark after Proposition 1.6.) By hypothesis, h is differentiable at some point $x_1 \in B$. A compactness argument shows that $h(x_1) = f(x_1, t_1) + g(t_1)$ for some $t_1 \in I$. This implies that $g(t_1) > -\infty$, so that t_1 is an interior point of I. We will use the characterization of Gâteaux differentiability as given in Exercise 1.24. For any $(y, s) \in E \times \mathbf{R}$ and all $t > 0$ sufficiently small, we have $(x_1 \pm ty, t_1 \pm ts) \in B \times I$ and

$$0 \le f(x_1 + ty, t_1 + ts) + f(x_1 - ty, t_1 - ts) - 2f(x_1, t_1) \le$$

$$h(x_1 + ty) + h(x_1 - ty) - 2h(x_1) - [g(t_1 + t) + g(t_1 - t) - 2g(t_1)].$$

Since both h and g are differentiable, if we divide through these inequalities by $t > 0$ and let $t \to 0$, the terms on the right tend to zero, showing that f is differentiable at (x_1, t_1).

Corollary 6.6. *A Banach space E is a GDS if and only if it is an MDS.*

Proof. Let H be a closed hyperplane in E, so that E is isomorphic to the product $H \times \mathbf{R}$. If E is an MDS, then by Prop. 6.4, H is a GDS. By Prop. 6.5, the space $H \times \mathbf{R}$, that is, the space E, is a GDS.

6.7. Problems.

Is a closed subspace of a GDS necessarily a GDS? Is the product of two Gâteaux differentiability spaces necessarily a GDS?

(It is obvious from Prop. 6.5 that, by induction, the second question has an affirmative answer for the product of a GDS with a finite dimensional space.)

The following permanence property for Gâteaux differentiability spaces is similar to the corresponding one for weak Asplund spaces (Theorem 4.19) which required the operator T to be *onto*; the proof is essentially the same.

Proposition 6.8. *Suppose that* $T: E \to F$ *is continuous and linear, with dense range. If the Banach space* E *is a GDS, then so is* F.

Proof. Suppose that D is a nonempty open convex subset of F and that f is continuous and convex on D. The function $f_1 = f \circ T$ is convex and continuous on the open convex set $D_1 = T^{-1}(D)$, hence is Gâteaux differentiable at the points of a dense set $G_1 \subset D_1$. To see that f is Gâteaux differentiable at the points of the dense set $T(G_1)$, it suffices to verify that $T^*[\partial f(Tx)] \subset \partial f_1(x)$ for all $x \in D_1$ and use the fact that density of $T(E)$ implies that T^* is one-one.

(It follows easily from the foregoing proposition that the first question in 6.7 has an affirmative answer for any *complemented* subspace M of a GDS space E; that is, for any M which is the image of a continuous linear projection on E.)

The next proposition is competely analogous to Prop. 5.11.

Proposition 6.9. *Suppose that* p *is a Minkowski gauge on* E. *An element* x^* *in* $C(p) = \{x^* \in E^*: \langle x^*, x \rangle \leq p(x) \text{ for all } x \in E\}$ *is weak* exposed by* $x \in E$ *if and only if* p *is Gâteaux differentiable at* x, *with* $dp(x) = x^*$.

Proof. Suppose that $x^* \in C$ and that x weak* exposes C at x^*; then $\partial p(x) = \{x^*\}$. Indeed, if $y^* \in \partial p(x)$, then by Lemma 5.10 we have $y^* \in C$ and $\langle y^*, x \rangle = p(x) = \sigma_C(x)$, hence $y^* = x^*$. Conversely, suppose that p is Gâteaux differentiable at x, with $x^* = dp(x)$. Then x^* is in $\partial p(x)$, so by Lemma 5.10 again, $x^* \in C$ and x attains its supremum on C at x^*. Suppose there were another point $y^* \in C$ such that $\sigma_C(x) = \langle y^*, x \rangle$; then the other implication in Lemma 5.10 shows that $y^* \in \partial p(x)$, hence $y^* = x^*$, that is, x weak* exposes C at x^*.

Recall the statement of Theorem 6.2: *A Banach space* E *is a Gâteaux differentiability space if and only if every weak* compact convex subset* C *of* E^* *is the weak* closed convex hull of its weak* exposed points.*

Proof of necessity in Theorem 6.2. The proof that the dual of a GDS has the indicated property is identical to the proof of the analogous portion of Theorem 5.12, except for the use of Prop. 6.9 in place of Prop. 5.11 and, of course, the substitution of Gâteaux differentiability for Fréchet differentiability. To prove the converse, we need the following dual version of a classical "parallel hyperplane lemma".

Lemma 6.10. *Suppose that $x, y \in E$, with $\|x\| = 1 = \|y\|$, and $\epsilon > 0$. If $|\langle x^*, y \rangle| \leq 1$ whenever $x^* \in E^*$ satisfies*

$$\langle x^*, x \rangle = 0 \text{ and } \|x^*\| < 2/\epsilon,$$

then either $\|x - y\| \leq \epsilon$ or $\|x + y\| \leq \epsilon$.

The following sketch shows the reason for the name; the hypotheses require that hyperplanes defined by the functionals x and y be nearly parallel.

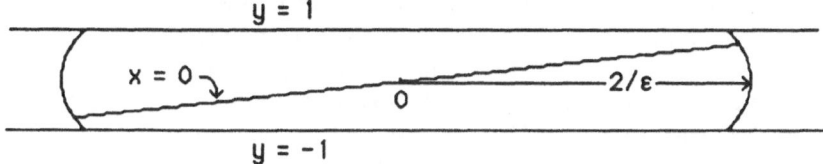

Fig. 6.2

Proof. Note that, as a (weak* continuous) linear functional on E^*, y is bounded in absolute value by $\epsilon/2$ on the intersection of the dual ball with the subspace $H = \{x^* : \langle x^*, x \rangle = 0\}$. By the Hahn-Banach theorem, its restriction to H can be extended to a functional of norm at most $\epsilon/2$ defined on all of E^*. Since this extension is necessarily weak* continuous, it is defined by an element $z \in E$. Thus, $y - z = 0$ on H, so $y - z = \alpha x$ for some $\alpha \in \mathbf{R}$. Note that

$$|1 - |\alpha|| = |\|y\| - \|y - z\|| \leq \|z\| \leq \epsilon/2.$$

If $\alpha \geq 0$, then $\|x - y\| = \|(1 - \alpha)x - z\| \leq |1 - \alpha| + \|z\| \leq \epsilon$.

If $\alpha < 0$, then $\|x + y\| = \|(1 + \alpha)x + z\| \leq |1 + \alpha| + \|z\| \leq \epsilon$.

Proof of sufficiency in Theorem 6.2. By Cor. 6.6, it suffices to show that E is an MDS. To that end, suppose that p is a Minkowski gauge functional on E, so that, by Lemma 5.10, we can write $p = \sigma_C$, where $C \equiv C(p)$ contains the origin. (This is the weak* compact convex set defined again in the statement of Prop. 6.9.) We lose no generality in assuming that $\sigma_C(x) \leq \|x\|$ for all $x \in E$. Since we want to show that σ_C (or, more simply, σ) is Gâteaux differentiable on a dense set and since it is clearly Gâteaux differentiable at each point of the interior of the closed convex set where σ vanishes, we need only consider the open set where σ is strictly positive. Suppose, then, that $x \in E$ and $\sigma(x) > 0$; we want to approximate x by a point y where σ is

Gâteaux differentiable. Since σ is positive homogeneous, we can assume that $\sigma(x) = 1$. Given $\epsilon > 0$ with $\epsilon < \sigma(x)$, it suffices to produce y such that $\|x - y\| < \epsilon$ and y exposes C. Consider the weak* compact convex subset C_1 of E^* defined by $C_1 = \text{co}(C \cup N)$, where

$$N = \{x^* \in E^* : \langle x^*, x \rangle = 0 \text{ and } \|x^*\| < 2/\epsilon\}.$$

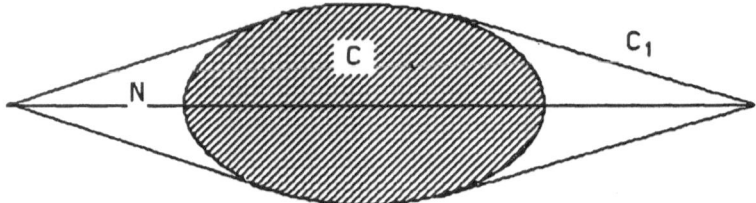

Fig. 6.3

Let σ_1 denote the support functional of C_1. By hypothesis, C_1 is the weak* closed convex hull of its weak* exposed points, so any weak* slice of C_1 contains a weak* exposed point. In particular, then, if $0 < \alpha \equiv \sigma(x) - \epsilon$, then the open slice $S(x, C_1, \alpha)$ contains a weak* exposed point y^* of C_1. Let $y \neq 0$ be a point of E which exposes C_1 at y^*; we may assume that $\|y\| = 1$. Since y^* is an extreme point of C_1 it is either in C or in N. The latter is impossible; since x vanishes on N we have $\sigma_1(x) = \sigma(x)$ so $\langle y^*, x \rangle > \sigma_1(x) - \alpha$ and the latter equals ϵ. Thus, y actually exposes C at y^*; since $0 \in C$ and $y^* \neq 0$, this implies in particular that $\langle y^*, y \rangle > 0$.

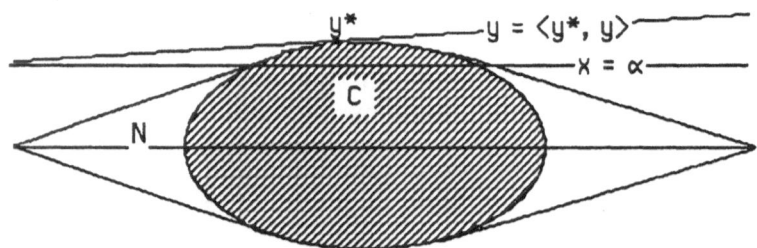

Fig. 6.4

The sketch above suggests that the hyperplanes defined by x and y are nearly parallel (hence that y is close to x). To see that this is actually the case, note first that since $\langle y^*, z \rangle \leq \sigma(z) \leq \|z\|$ for all z, we must have $\|y^*\| \leq 1$, hence $\langle y^*, y \rangle \leq 1$. Thus, if $x^* \in N \subset C_1$, then

$$\langle x^*, y \rangle \leq \sigma_1(y) = \langle y^*, y \rangle \leq 1;$$

by symmetry of N this implies that $|\langle x^*, y \rangle| \leq 1$ for all $x^* \in N$. It follows from Lemma 6.10 that either $\|x - y\| \leq \epsilon$ or $\|x + y\| \leq \epsilon$. But the latter is impossible: $\|y^*\| \leq 1$ hence

$$\|x + y\| \geq \langle y^*, x + y \rangle > \langle y^*, x \rangle > \epsilon.$$

Remarks.

Much of the material in this section originated in [La-Ph], with one major exception. Prop. 6.5 was shown to us by M. Fabian, who says the proof was motivated by work of A. Ioffe. (The function g introduced in the proof is what optimization specialists call a "penalty" function.) Fabian's result represents the first bit of affirmative progress in this topic in the last decade, although there have been a number of relevant striking examples (see [Č-K], [Tal₁] and [Tal₂]). We are grateful for his permission to include it in these notes. The names "Asplund space" and "weak Asplund space" were introduced in [Na-Ph]; Asplund called them "strong" and "weak differentiability spaces". In retrospect, both definitions could easily have dropped the G_δ requirement; indeed, it wouldn't have changed anything in the Fréchet case, and in the Gâteaux case, Asplund's name would now be attached to what appears to be a more tractable class of spaces.

J. Borwein and S. Fitzpatrick [Bor-F₂] have sketched a way to unify the results of this section with some of those of Sec. 5. A key (and easily proved) observation is the following: An element x^* of a weak* compact convex subset $C \subset E^*$ is weak* exposed by $x \in E$ provided $\sigma_C(x) = \langle x^*, x \rangle$ and $x_n^* \to x^*$ in the weak* topology whenever $\{x_n^*\} \subset C$ and $\langle x_n^*, x \rangle \to \langle x^*, x \rangle$. Now, the weak* topology is precisely the topology (call it G^o) of uniform convergence on elements of the Gâteaux bornology $\beta = G$ and the norm topology in E^* is the topology F^o of uniform convergence on the elements of the Fréchet bornology $\beta = F$. One is thus led to a unified definition of a *weak* β-exposed point* of such a set C: replace "weak* convergence" in the observation above by "convergence in the topology β^o". One can also define a "β-differentiability space" to be one in which every convex continuous function is β-differentiable on a dense subset of its domain. This makes it possible to mimic the proofs of Theorems 5.12 and 6.2 to prove the general result that *a Banach space E is a β-differentiability space if and only if every weak* compact convex subset of E^* is the weak* closed convex hull of its weak* β-exposed points.*

Section 7

7. A generalization of monotone operators: Usco maps.

There has been much relatively recent work showing that Kenderov's results about generic continuity of maximal monotone operators are special cases of theorems of a more topological nature. We will describe some of this, selecting the results most closely related to our primary interest in Banach spaces.

Definition 7.1. (a) Suppose that X and Y are Hausdorff topological spaces and let $F : X \to 2^Y$ be a map from X into the nonempty compact subsets of Y. We say that F is *usco* (for *upper semicontinuous and compact valued*) provided it is also upper semicontinuous. (Recall that upper semicontinuity means that $\{x \in X : F(x) \subset U\}$ is open in X whenever U is open in Y.)

(b) If Y is a Hausdorff linear space, then a usco map $F : X \to 2^Y$ is called *convex* usco provided $F(x)$ is convex for each $x \in X$.

(c) We denote the graph of F by $G(F) \equiv \{(x, y) \in X \times Y : y \in F(x)\}$, and, as with monotone operators, we partially order these set-valued maps by the inclusion ordering on their graphs. We will usually write $F_1 \subset F_2$ in place of $G(F_1) \subset G(F_2)$, and talk about one such map containing another. A usco map [convex usco map] is said to be *minimal* if it does not properly contain any other usco [convex usco] map.

(d) If $Y = E^*$ is a dual Banach space, saying that a usco map F from X into 2^Y is *w^*-usco* means that we are taking Y to be E^* in its weak* topology (so the values of F are weak* compact sets).

7.2. Example.

Suppose that E is a Banach space, that $T : E \to 2^{E^*}$ is maximal monotone and that D is a nonempty open subset of $D(T)$. Then the restriction T_D of T to D is convex w^*-usco (as a set-valued map from D to E^* in its weak* topology).

Proof. This is a consequence of Exercise 2.29.

The kind of theorem we will consider has the form: For certain minimal usco maps F from a Baire space X into 2^Y, there exists a dense G_δ subset D of X such that F is single-valued and continuous (for an appropriate topology in Y) at each point of D. (This is not too surprising, since we would expect minimal set-valued maps to be single-valued at many points.) In view of the relationship between of Asplund spaces E and certain continuity properties of maximal monotone operators defined in E, it is also not surprising that such results would be of interest. We need some elementary facts about usco maps.

Proposition 7.3. *(a) If $F: X \to 2^Y$ is usco, then its graph $G(F)$ is closed in $X \times Y$.*

(b) For every usco map [convex usco map] F there exists a minimal usco map [minimal convex usco map] contained in F.

Proof. (a) Suppose that (x_α, y_α) is a net in $G(F)$ which converges to a point $(x, y) \in X \times Y$, but that y is not in $F(x)$. Since the latter is compact there exists an open neighborhood U of $F(x)$ whose closure $\overline{U}$ does not contain y. By upper semicontinuity, (y_α) is eventually in U, and hence y is in $\overline{U}$, a contradiction.

(b) This will follow from Zorn's lemma provided we can show that any decreasing chain $\{F_\alpha\}$ of [convex] usco maps contained in F has a minimal element. For $x \in X$ define $F_0(x) = \cap F_\alpha(x)$; by compactness, this is nonempty, [convex] and compact. To see that F_0 is upper semicontinuous, suppose that $x \in X$, U is open in Y and $F_0(x) \subset U$. Necessarily, $F_\alpha(x) \subset U$ for some α; indeed, if each $F_\alpha(x) \setminus U$ were nonempty, the intersection of these compact nested sets would be a nonempty subset of $F_0(x) \setminus U$. By upper semicontinuity of F_α, there exists an open set V containing x such that $F_0(V) \subset F_\alpha(V) \subset U$.

Example 7.2 showed that maximal monotone operators are convex usco maps; the following example shows that they need not be minimal usco maps.

7.4 Example.

Define monotone operators F_0, F_1 and F_2 from $\mathbf{R}$ to $2^{\mathbf{R}}$ by

$$F_0(x) = F_1(x) = F_2(x) = \text{sgn } x \text{ if } x \neq 0,$$

while

$$F_0(0) = \{-1\}, F_1(0) = \{-1, 1\} \text{ and } F_2(0) = [-1, 1].$$

Then $F_0 \subset F_1 \subset F_2$, they are all distinct, F_2 is maximal monotone and minimal convex usco, F_1 is minimal usco (hence has closed graph) and F_0 does not have closed graph, although $F_0(x)$ is closed for each x.

Proof. Since $F_2 = \partial f$, where $f(x) = |x|$ is continuous and convex, F_2 is maximal monotone. The remaining assertions are readily verified.

The following sufficient condition for a map to be usco is quite useful.

Proposition 7.5. *Suppose that F_1, F_2 are set-valued maps from X into Y such that $G(F_2)$ is closed, $F_2 \subset F_1$ and F_1 is usco. Then F_2 is usco.*

Proof. The fact that $G(F_2)$ is closed implies that $F_2(x)$ is closed for each $x \in X$. Since $G(F_2) \subset G(F_1)$ implies that $F_2(x) \subset F_1(x)$, each $F_2(x)$ is necessarily compact. It remains to prove that F_2 is upper semicontinuous. Suppose that $x \in X$ but that F_2 is not upper semicontinuous at x. Then there exist an open set $U \subset Y$ containing $F_2(x)$, a net $(x_\alpha) \subset X$ such that $x_\alpha \to x$, and points $y_\alpha \in F_2(x_\alpha) \setminus U$ for each α. Since F_1 is upper semicontinuous at x, the net (y_α) is eventually in every open neighborhood of $F_1(x)$. This implies that some cluster point y of (y_α) is in the compact set $F_1(x)$. (Indeed, if not, then for every $y \in F_2(x)$ there would exist an open neighborhood V_y of y such that (y_α) is eventually in $Y \setminus V_y$. By covering $F_1(x)$ with such neighborhoods and using compactness, we could find an open set W containing $F_1(x)$ such that (y_α) is eventually outside W.) Take a subnet of (x_α, y_α) converging to (x, y); since $G(F_2)$ is closed, we conclude that $y \in F_2(x) \subset U$. On the other hand, $y_\alpha \in Y \setminus U$ for all α, so $y \in Y \setminus U$, a contradiction which completes the proof.

We want to extend the notion of "maximal monotone" in the following way:

Definition 7.6. If $T: E \to 2^{E^*}$ is monotone and $D \subset E$, we say that T is *maximal monotone in D* provided the monotone set

$$G(T) \cap (D \times E^*) \equiv \{(x, x^*) \in D \times E^*: x \in D \text{ and } x^* \in T(x)\}$$

is maximal (under set inclusion) in the family of all monotone sets contained in $D \times E^*$.

We will see shortly (Cor. 7.8, below) that if T is a maximal monotone operator in E, then its restriction to any open subset D of $D(T)$ is maximal monotone in D.

Lemma 7.7. *Suppose that $D \subset E$ is open, that $T: D \to 2^{E^*}$ is monotone and norm-to-weak* upper semicontinous and that $T(x)$ is nonempty, convex and weak* closed for all $x \in D$; then T is maximal monotone in D.*

Proof. As with the usual notion of maximal monotone, we need only show that if $(y, y^*) \in D \times E^*$ satisfies

$$\langle y^* - x^*, y - x \rangle \geq 0 \text{ for all } x \in D, x^* \in T(x), \tag{7.1}$$

then $y^* \in F(y)$. If not, by the separation theorem there exists $z \in E$ such that $F(y) \subset \{z^* \in E^*: \langle z^*, z \rangle < \langle y^*, z \rangle\} = W$. Since W is weak* open and T is norm-to-weak* upper semicontinuous, there exists a neighborhood U of y in D such that $T(U) \subset W$. For $t > 0$ sufficiently small, $y + tz \in U$ and therefore $T(y + tz) \subset W$. Applying (7.1) to any $u^* \in T(y + tz)$ we get

$$0 \leq \langle y^* - u^*, y - (y + tz) \rangle = -t\langle y^* - u^*, z \rangle,$$

which implies that $\langle u^*, z \rangle \geq \langle y^*, z \rangle$, that is, u^* is not in W, a contradiction.

Corollary 7.8. *If $T: E \to 2^{E^*}$ is maximal monotone and D is a nonempty open subset of $D(T)$, then the restriction T_D of T to D is maximal monotone in D.*

Proof. By Example 7.2, the maximal monotonicty of T implies that the monotone map T_D is convex w^*-usco, so the result follows from Lemma 7.7.

The next theorem exhibits an interesting relationship between maximal monotone operators in an open set D and minimal usco maps in D.

Theorem 7.9. *If D is open in the Banach space E and if $T: D \to 2^{E^*}$ has nonempty values and is maximal monotone in D, then T is a minimal convex w^*-usco map in D.*

Proof. We know by Example 7.2 that T is convex and w^*-usco, so the only question is whether it is minimal in this family. Suppose that $F: D \to 2^{E^*}$ is convex and w^*-usco and that $G(F) \subset G(T)$. By Lemma 7.7, F is maximal monotone and hence $F = T$.

It is easy to see that not every minimal convex w^*-usco is monotone; consider any continuous non-monotonic function from **R** into itself.

The application of these notions (usco maps, minimal usco maps, etc.) to a generic continuity theorem for monotone operators requires several further lemmas.

Lemma 7.10. *Let X, Y be Hausdorff spaces and suppose that $F: X \to 2^Y$ has closed graph and is locally relatively compact, that is, every $x \in X$ has an open neighborhood V such that $F(V)$ is relatively compact in Y. Then F is a usco map.*

Proof. Since $G(F)$ is closed, every image $F(x)$ is closed and contained in some relatively compact set, hence is compact. To see that F is upper semicontinuous, it suffices to show that for each x in X and each open neighborhood V of x, the restriction of F to V is upper semicontinuous. By hypothesis, we may restrict attention to open neighborhoods V of x for which $\overline{F(V)}$ is compact. Given such a V, define $F_1: V \to 2^Y$ by $F_1(x) = \overline{F(V)}$, $x \in V$. Obviously, F_1 is a usco map and $F|_V \subset F_1$; since $G(F|_V)$ is closed in $V \times Y$, Proposition 7.5 implies that F is upper semicontinuous in V.

Lemma 7.11. *If D is open in E and $T: D \to 2^{E^*}$ is monotone, with $T(x)$ nonempty for each $x \in D$, then the map $\overline{T}$ whose graph is the closure $\overline{G(T)}$ in $D \times (E^*, w^*)$ of $G(T)$ is monotone and w^*-usco.*

Proof. Let $T_1: D \to 2^{E^*}$ be a maximal monotone extension of T. By Theorem 7.9, T_1 is a minimal convex w^*-usco map in D, and by Prop. 7.3 its graph $G(T_1)$ is closed in $D \times (E^*, w^*)$. Thus, $\overline{T} \subset T_1$ so $\overline{T}$ is monotone and—by Prop. 7.5—it is also a w^*-usco map.

The next theorem shows how the concept of minimal usco maps can be used to prove a basic result about monotone operators.

Lemma 7.12. *Suppose that X is a Hausdorff space and that $T: X \to 2^{E^*}$ is w^*-usco. For $x \in X$ define $\overline{co}T(x)$ to be the weak* closed convex hull of $T(x)$. Then the map $\overline{co}T$ is convex w^*-usco.*

Proof. Since $\overline{co}T$ obviously has weak* compact convex values, it suffices to prove that it is weak* upper semicontinuous. To see this, suppose $x \in X$ and that U is a weak* open subset of E^* with $\overline{co}T(x) \subset U$. In any locally convex space, a compact convex set K has a neighborhood base of the form $K + W$, where the closed convex sets W form a neighborhood base of 0. Thus, we can assume that U is of the form $U = \overline{co}T(x) + W$, where W is a weak* closed convex neighborhood of 0. By the upper semicontinuity of T there exists an open neighborhood V of x in X such that

$$T(V) \subset \overline{co}T(x) + W.$$

It follows that $\overline{co}T(V) \subset \overline{co}T(x) + W$, so $\overline{co}T$ is upper semicontinuous.

Theorem 7.13. *Suppose that D is an open subset of the Banach space E and that $T: D \to 2^{E^*}$ is a monotone operator, with $T(x) \neq \emptyset$ for all x in D. Then there is a unique maximal monotone operator M in D containing T. In fact, M can be characterized as follows: Let $\overline{T}$ be the set-valued map whose graph is the closure in $D \times (E^*, w^*)$ of $G(T)$ and for each $x \in D$ let $M(x)$ be the weak* closed convex hull of $\overline{T}(x)$; this defines M.*

Remark. As Example 7.4 shows, one <u>must</u> distinguish between sets of the form $\overline{T}(x)$ and the (possibly smaller) sets $\overline{T(x)}$.

Proof. Let T_1 be any maximal monotone operator containing T. By Theorem 7.9, T_1 is a minimal convex w^*-usco map. Since it has closed graph, we must have $\overline{T} \subset T_1$, and since $\overline{T}$ has closed graph, Prop. 7.5 implies that it is w^*-usco. From Lemma 7.12, we conclude that the map $M = \overline{co}\overline{T}$ is convex w^*-usco and, clearly, $M \subset T_1$. By the minimality of T_1, we have $M = T_1$, which proves the uniqueness assertion.

There does not appear to be a unique extension theorem for monotone operators with arbitrary effective domains. If, for instance, E has dimension at least one, if $(x_0, y_0) \in E \times E^*$ and if T is defined to be the monotone operator whose graph is $\{(x_0, y_0)\}$, then there are many maximal monotone extensions of T.

The next lemma, which is purely topological in nature, has Kenderov's Theorem 2.30 on continuity of maximal monotone mappings as an immediate corollary. The main hypothesis will seem less peculiar when we apply the lemma to maximal monotone operators.

Lemma 7.14. *Let F be a minimal usco map on the Baire space X with compact values in the Hausdorff space (Y, τ) and let d be a metric on Y. Suppose that for every nonempty open subset U of X there exist nonempty open* subsets V of U such that $F(V)$ contains relatively open subsets of arbitrarily small d-diameter. Then there exists a dense G_δ subset D of X such that F is single-valued and d-upper semicontinuous at each point of D.

Proof. We first note that the fact that F is a minimal usco map implies that if J is a proper closed subset of $G(F)$, then $p(J) \neq X$, where p is the natural projection of $X \times Y$ onto X. Indeed, if $p(J) = X$, then J would be the (closed) graph of a set-valued map which, by Prop. 7.5, would be a usco map properly contained in F. Next, given $\epsilon > 0$, let

$$O_\epsilon = \cup\{G: G \text{ is an open subset of } X \text{ and } d - \text{diam } F(G) \leq \epsilon\}.$$

Clearly, O_ϵ is open in X; we will show that it is dense. Let U be a nonempty open subset of X. By hypothesis, there is a nonempty open subset V of U and a τ-open subset W of Y such that $F(V) \cap W \neq \emptyset$ and d-diam $(F(V) \cap W) \leq \epsilon$. Since $G(F) \cap (V - W) \neq \emptyset$ and F is minimal (and, by Prop. 7.3 (a), $G(F)$ is closed), we must have $p[G(F) \setminus (V \times W)] \neq X$. Choose $x_0 \in X \setminus p[G(F) \setminus (V \times W)]$. Then

$$p^{-1}(x_0) \cap G(F) \subset V \times W,$$

that is, $x_0 \in V$ and $F(x_0) \subset W$. If $G = \{x \in X: F(x) \subset W\} \cap V$, then G is an open neighborhood of x_0 with d-diam $F(G) \leq d$-diam$(F(V) \cap W) \leq \epsilon$. It follows that $G \subset O_\epsilon$ and therefore $\emptyset \neq V \cap O_\epsilon \subset U \cap O_\epsilon$. This proves that O_ϵ is dense in X. Now let

$$D = \cap\{O_{1/n}: n = 1, 2, 3, ...\}.$$

Since X is a Baire space, D is a dense G_δ subset of X. From the definition of D it is evident that not only is $F(x)$ a singleton at each point $x \in D$ but that F is d-upper semicontinuous at each such point.

This lemma leads to the following alternative proof of Theorem 2.30.

Theorem 7.15. *Let E be a Banach space such that every bounded nonempty subset of E^* is weak* dentable. If $T: E \to 2^{E^*}$ is maximal monotone, with $X \equiv \text{int } D(T) \neq \emptyset$, then there exists a dense G_δ subset D of X such that T is single-valued and norm-to-norm upper semicontinuous at each point of D.*

Proof. We know from Example 7.2, Cor. 7.8 and Theorem 7.9 that T is a minimal convex w^*-usco map from X into (E^*, w^*); by Prop. 7.3 it contains a minimal w^*-usco map T_1. By Theorem 2.28, T is locally bounded in X, hence the same is true of T_1. Consequently, given any nonempty open subset U of X there exists a nonempty open subset V of U such that $T_1(V)$ is bounded. By the weak* dentability hypothesis, $T_1(V)$ admits nonempty relatively weak* open neighborhoods of arbitrarily small norm diameter, so by Lemma 7.14 there exists a dense G_δ subset D of X such that T_1 is single-valued and norm-to-norm upper semicontinuous at each point of D. Now, by Lemma 7.12, $\overline{co}T_1$ is convex w^*-usco, and hence by the minimality of $T|_X$, we have $\overline{co}T_1 = T|_X$.

Thus, for $x \in X$, if $T_1(x)$ is contained in a closed ball, then so is $T(x)$. It follows that T is also single-valued and norm-to-norm upper semicontinuous at each $x \in D$.

The following exercise exhibits a particular convex w*–usco map which is of considerable importance in optimization.

7.16. Exercises: The Clarke subdifferential.

Suppose that f is a locally Lipschitzian real-valued function on a nonempty open subset D (not necessarily convex) of the Banach space E. For each $x \in D$ the *generalized directional derivative* $f^\circ(x; h)$ of f at x in the direction $h \in E$ is defined by

$$f^\circ(x; h) = \lim_{(t,y)\to(0^+,x)} \sup \frac{f(y + th) - f(y)}{t}.$$

(a) Show that $f^\circ(x; h)$ is finite and that for fixed x, the function $h \to f^\circ(x; h)$ is positive homogeneous, subadditive and Lipschitzian.

(b) Show that if $\{x_n\} \subset D$ and $x_n \to x \in D$, then for each $h \in E$,

$$\lim \sup f^\circ(x_n; h) \le f^\circ(x; h).$$

The *Clarke subdifferential* $\partial^\circ f(x)$ of f at $x \in D$ is defined by

$$\partial^\circ f(x) = \{x^* \in E^* : \langle x^*, h \rangle \le f^\circ(x; h) \quad \text{for all } h \in E\}.$$

(c) Show that, for all $x \in E$, the set $\partial^\circ f(x)$ is a nonempty convex weak* compact subset of E^*.

(d) Show that the graph of $\partial^\circ f$ is norm $\times$ weak* closed in $D \times E^*$ and hence that $\partial^\circ f$ is a convex weak* usco map whenever $D = E$.

If f is a continuous convex function, then it is locally Lipschitzian, hence one can define both its generalized directional derivative and Clarke subdifferential as above. The following proposition reassures us that these coincide with the usual notions of directional derivative and subdifferential in this case.

Proposition 7.16. *If f is convex and continuous on a nonempty open convex subset D of E (hence is locally Lipschitzian), then for all $x \in D$ and $h \in E$ one has*

$$f^\circ(x; h) = d^+ f(x)(h)$$

and hence $\partial^\circ f(x) = \partial f(x)$.

Proof. It is immediate from the definitions that $f^\circ(x; h) \ge d^+ f(x)(h)$. To prove the reverse inequality, note that for any fixed $\delta > 0$,

$$f^\circ(x; h) = \lim_{\substack{\epsilon\to 0^+ \\ \|y-x\|<\delta \\ 0<t<\epsilon}} \sup \frac{f(y + th) - f(y)}{t}. \qquad (7.2)$$

Since (as in Lemma 1.2) the difference quotient

$$\frac{f(y + th) - f(y)}{t}$$

is nonincreasing as $t \to 0^+$, the right side of (7.2) equals

$$\lim_{\epsilon \to 0^+} \sup_{\|y - x\| < \epsilon\delta} \frac{f(y + \epsilon h) - f(y)}{\epsilon}.$$

Now, there exists a neighborhood of x in which f has Lipschitz constant $M > 0$, say, so for all sufficiently small $\epsilon > 0$, if $\|x - y\| < \epsilon\delta$, then

$$\frac{f(y + \epsilon h) - f(y)}{\epsilon} \leq \frac{f(x + \epsilon h) - f(x)}{\epsilon} + 2\delta M$$

and therefore $f^\circ(x; h) \leq d^+ f(x)(h) + 2\delta M$. This being true for all $\delta > 0$, we obtain the desired inequality. The fact that the two subdifferentials coincide follows from the foregoing equality, the definition of $\partial^\circ f(x)$ and the fact (shown in the proof of Prop. 1.8) that $\partial f(x) = \{x^* \in E^* : \langle x^*, h \rangle \leq d^+ f(x)(h) \quad \text{for all } h \in E\}$.

Theorem 7.9 shows that if a monotone operator which takes on nonempty values is an open set D and is maximal monotone in D is necessarily a *minimal* convex w*-usco in D. Although the Clarke subdifferential of a Lipschitzian function f on an open set D is a convex w*-usco in D, it need not be a minimal convex w*-usco in D. Of course, it is minimal if f is convex, and J. Borwein [Bor$_2$] has shown that it is minimal for certain other related classes of functions.

Remarks.

Most of the results in this section were obtained independently by Drewnowski and Labuda [Dr-L] and Jokl [Jo]; we have followed the exposition of [Dr-L], with some very significant improvements suggested by both S. Fitzpatrick and I. Namioka. The proof of Lemma 7.14 (which results in the new proof of one of Kenderov's theorems in Theorem 7.15) is a minor revision of one due to the highly esteemed John Rainwater [Rain], who was motivated by the work of M. E. Verona [Ver]. Between them, they have extended some of the generic differentiability results to Lipschitz continuous convex functions defined on a sort of quasi-interior of a closed convex set C (namely the Baire space of all non-support points of C). Some of their theorems were obtained independently by D. Noll [Noll$_{1,2}$]. Far-reaching generalizations of these kinds of results have been obtained by J. Borwein, S. Fitzpatrick and P. Kenderov [B-F-K] and by M. E. Verona and A. Verona [V-V$_{1,2,3}$].

As mentioned above, the Clarke subdifferential plays an important role in optimization; see, for instance, [Cl] for further properties and applications. While it is a very nice extension to locally Lipschitzian functions of the usual notion for convex functions, it has some drawbacks. For instance, it does not apply to lower semicontinuous convex functions which are not continuous (hence not locally Lipschitzian), and it fails to be single-valued for some differentiable functions (such as $f(x) = x^2 \sin\frac{1}{x}$ for $x \neq 0$, $f(0) = 0$; in this case it is easy to verify that $\partial^\circ f(0) = [-1, 1]$).

References

[A-L]   D. Amir and J. Lindenstrauss, The structure of weakly compact subsets in Banach spaces, Ann. Math. 88 (1968), 35 - 46.

[Asp]   E. Asplund, Fréchet differentiability of convex functions, Acta Math.121 (1968), 31-47.

[Au-Ek]   J-P Aubin and I. Ekeland, Applied Nonlinear Analysis, Wiley Interscience, New York (1984).

[Bi-Ph]   E. Bishop and R. R. Phelps, The support functionals of a convex set, Proc. Symp. Pure Math. Vol. 7, Amer. Math. Soc. (1962), 27-35

[B-F-P]   J. Borwein, S. Fitzpatrick and P. Kenderov, Minimal convex uscos and monotone operators on small sets, Canad. J. Math. 43 (1991), 461-476.

[Bor$_1$] J. M. Borwein, A note on ϵ-subgradients and maximal monotonicity, Pac. J. Math. 103 (1982), 307-314.

[Bor$_2$] _____, Minimal cuscos and subgradients of Lipschitz functions, Fixed Point Theory and its Applications (J.-B. Baillon and M. Thera, eds), Pitman Lecture Notes in Math. , Longman, Essex (1991), 57-82.

[Bor$_3$] _____, Asplund spaces are "sequentially reflexive", Canad. J. Math. (to appear)

[Bor-F$_1$] J. Borwein and S. P. Fitzpatrick, Local boundedness of monotone operators under minimal hypotheses, Bull. Australian Math. Soc. 39 (1989), 439-441.

[Bor-F$_2$] _____, A weak Hadamard smooth renorming of $L_1(\Omega, \mu)$, Canad. Math. Bull. (to appear)

[Bor-P]   J. M. Borwein and D. Preiss, A smooth variational principle with applications to subdifferentiability and to differentiability of convex functions, Trans. Amer. Math. Soc. 303 (1987), 517-527.

[Bo$_1$]   J. Bourgain, On dentability and the Bishop-Phelps property, Israel J. Math., 28 (1977), 265-271.

[Bo$_2$]   _____, La propriété de Radon-Nikodým, Publ. Math. de l'Univ. Pierre et Marie Curie, Nr. 36 (1979).

[Bo-Ta]   J. Bourgain and M. Talagrand, Dans un espace de Banach reticulé solide, la propriété de Radon-Nikodym et celle de Krein-Milman sont équivalentes, Proc. Amer. Math. Soc. 81 (1981), 93-96.

[Bou]   R. D. Bourgin, Geometric aspects of convex sets with the Radon-Nikodym property, Lect. Notes in Math., Nr. 993, Springer-Verlag (1983).

[Bre]   H. Brezis, Opérateurs Maximaux Monotones et semi-groupes de contractions dans les espaces de Hilbert, Math. Studies 5, North-Holland American Elsevier (1973).

[Br]   F. E. Browder, Multivalued monotone nonlinear mappings and duality mappings in Banach space, Trans. Amer. Math. Soc. 118 (1965), 338-351.

[Ca]   V. Caselles, A short proof of the equivalence of KMP and RNP in Banach lattices and preduals of von Neumann algebras, Proc. Amer. Math. Soc. 102 (1988), 973-974.

[Ch]   G. Choquet, Lectures on Analysis, vol. I, W. A. Benjamin, New York (1969).

[Chr]   J. P. R. Christensen, Theorems of Namioka and R. E. Johnson type for upper semicontinuous and compact-valued set-valued mappings, Proc. Amer. Math Soc. 86 (1982), 649-655.

[Chr-K]   J. P. R. Christensen and P. S. Kenderov, Dense strong continuity of mappings and the Radon-Nikodym property, Math. Scand. 54 (1984), 70-78.

[Chu]   C.-H. Chu, A note on scattered C*-algebras and the Radon-Nikodym property, J. London Math. Soc. (2) 24 (1981), 533-536.

[C-K]   M. Coban and P. S. Kenderov, Dense Gâteaux differentiability of the sup-norm in $C(T)$ and the topological properties of T, C. R. Acad. Bulgare Sci. 38 (1985), 1603-1604.

[Co]   J. B. Collier, The dual of a space with the Radon-Nikodym property, Pacific J. Math. 64 (1976), 103-106.

[Cr-Li$_1$] M. Crandall and P.-L. Lions, Hamilton-Jacobi equations in infinite dimensions I. Uniqueness of viscosity solutions, J. Funct. Analysis 62 (1985), 379-396.

[Cr-Li$_2$] _____, Hamilton-Jacobi equations in infinite dimensions II. Existence of viscosity solutions, ibid. 65 (1986), 368-405.

[De]   K. Deimling, Nonlinear Functional Analysis, Springer-Verlag (1985).

[D-G-Z$_1$] R. Deville, G. Godefroy and V. Zizler, Un principe variationnel utilisant des fonctions bosses, C. R. Acad. Sci. Paris 312, Série I, (1991), 281-286.

[D-G-Z$_2$] _____, Renormings and Smoothness in Banach Spaces, Monographs and Surveys in Pure and Appl. Math., Longman (to appear)

[D-G-Z$_3$] _____, A smooth variational principle with applications to Hamilton-Jacobi equations in infinite dimensions, J. Functional Anal. 111 (1993), 197-212,

[Di]   J. Diestel, Geometry of Banach space—selected topics, Lect. Notes in Math., Nr. 485, Springer-Verlag (1975).

[Di-U]   J. Diestel and J. J. Uhl, Jr., Vector Measures, Math. Surveys 15, Amer. Math. Soc. (1977).

[Du-N]   D. v. Dulst and I. Namioka, A note on trees in conjugate Banach spaces, Indag. Math. 46 (1984), 7-10.

[Dr-L$_1$] L. Drewnowski and I. Labuda, On minimal convex usco and maximal monotone maps, Real Analysis Exchange 15 (1989-90), 729-742.

[Dr-L$_2$] _____, On minimal upper semicontinuous compact-valued maps, Rocky Mountain J. Math. 20 (1990), 737-752.

[Ed-W]   G. A. Edgar and R. F. Wheeler, Topological properties of Banach spaces, Pacific J. Math. 115 (1984), 317-350.

[Ek]   I. Ekeland, Nonconvex minimization problems, Bull. Amer. Math. Soc. (New Series) 1 (1979), 443-474.

[Ek-L]   I. Ekeland and G. Lebourg, Generic Fréchet differentiability and perturbed optimization problems in Banach spaces, Trans. Amer. Math. Soc. 224 (1976), 193-216.

[Ek-T]   I. Ekeland and R. Temam, Convex Analysis and Variational Problems, Studies in Math. and its applications, North-Holland American Elsevier, New York (1976).

[Fa$_1$]   M. Fabian, Every weakly countably determined Asplund space admits a Fréchet differentiable norm, Bull. Austr. Math. Soc. 36 (1987), 367-374.

[Fa$_2$]   _____, On minimum principles, Acta Polytechnica 20 (1983), 109-118.

[Fi$_1$]   S. P. Fitzpatrick, Monotone operators and dentability, Bull. Australian Math. Soc. 18 (1978), 77-82.

[Fi$_2$]   _____, Representing monotone operators by convex functions, Proc. Centre for Math. Analysis 20 (1989), 59-65.

[Fl]   T. M. Flett, Differential Analysis, Cambridge University Press, Cambridge (1980).

[G-L-M]   N. Ghoussoub, J. Lindenstrauss and B. Maurey, Analytic martingales and plurisubharmonic barriers in complex Banach spaces, Contemp. Math. 85 (1989), 111-130.

[Gh-M₁]   N. Ghoussoub and B. Maurey, G_δ-embeddings in Hilbert space, J. Funct. Analysis 61 (1985), 72-97.

[Gh-M₂]   —————, H_δ-embeddings in Hilbert space and optimization on G_δ-sets, Memoirs A. M. S., Nr. 349 (1986).

[Gi]   J. R. Giles, Convex analysis with application in differentiation of convex functions, Res. Notes in Math., Nr. 58, Pitman, Boston-London-Melbourne, (1982).

[Ha₁]   R. Haydon, A counterexample to several questions about scattered compact spaces, Bull. London Math. Soc. 22 (1990), 261-268.

[Ha₂]   —————, Trees in renorming theory, (preprint)

[Jo]   L. Jokl, Minimal convex-valued weak* usco correspondences and the Radon-Nikodym property, Comm. Math. Univ. Carolinae 28 (1987), 353-375.

[Ke₁]   P. S. Kenderov, The set-valued monotone mappings are almost everywhere single-valued, C. R. Acad. Bulgare Sci. 27 (1974), 1173-1175.

[Ke₂]   —————, Monotone operators in Asplund spaces, C. R. Acad. Bulgare Sci.30 (1977), 963-964.

[Kl]   V. Klee, Some new results on smoothness and rotundity in normed linear spaces, Math. Ann. 139 (1959), 51-63.

[Kr₁]   E. Krauss, A representation of maximal monotone operators by saddle functions, Rev. Roum. Math. Pures Appl. 30 (1985), 823-837.

[Kr₂]   —————, Maximal monotone operators and saddle functions I, Zeitschr. für Anal. u. ihre Anw. 5 (1986), 336-346.

[La-Ph]   D. G. Larman and R. R. Phelps, Gâteaux differentiability of convex functions on Banach spaces, J. London Math. Soc. 20 (1979), 115-127.

[Li]   J. Lindenstrauss, On operators which attain their norm, Israel J. Math. 1 (1963), 139-148.

[Ma]   S. Mazur, Über konvexe Mengen in linearen normierten Räumen, Studia Math. 4 (1933), 70-84.

[Na-Ph]   I. Namioka and R. R. Phelps, Banach spaces which are Asplund spaces, Duke Math. J. 42 (1975), 735-750.

[Noll₁]   D. Noll, Generic Gâteaux differentiability of convex functions on small sets, J. Math. Analysis and Appl. 147 (1990), 531-544.

[Noll₂]   —————, Generic Fréchet differentiability of convex functions on small sets, Arch. Math. (54) (1990), 487-492.

[Ør]   P. Ørno, On J. Borwein's concept of sequentially reflexive Banach spaces, (TeX electronic manuscript) File:pub/banach/orno.tex. Banach space bulletin board archive: ftp.math.okstate.edu. Posted 10-9-91.

[Ox]   J. O. Oxtoby, The Banach-Mazur game and Banach category theorem, in Contributions to the Theory of Games, vol III, Annals of Math. Studies 39, Princeton, N. J. (1957), 159-163.

[Pa-Sb]   D. Pascali and S. Sburlan, Nonlinear mappings of monotone type, Ed. Acad., Bucarest, Rom., Sijthoff & Noordhoff Internat. Publ., Alphen aan den Rijn, Netherlands (1978).

[Ph₁]   R. R. Phelps, A representation theorem for bounded convex sets, Proc. Amer. Math. Soc. 11 (1960), 976-983.

[Ph₂]   —————, Dentability and extreme points in Banach spaces, J. Functional Analysis 17 (1974), 78 -90.

[Ph₃] —————————, Convexity in Banach spaces: some recent results, Convexity and its Applications, Gruber and Wills, Ed., Birkhäuser Verlag, Basel-Boston-Stuttgart (1983), 277-295.

[Ph₄] —————————, Convex Functions, Monotone Operators and Differentiability, Lect. Notes in Math., Nr. 1364, Springer-Verlag (1989).

[Pr] D. Preiss, Fréchet derivatives of Lipschitz functions, J. Funct. Analysis 91 (1990), 312-345.

[P-P-N] D. Preiss, R. R. Phelps and I. Namioka, Smooth Banach spaces, weak Asplund spaces and monotone or usco mappings, Israel J. Math. 72 (1990), 257-279.

[Pr-Z] D. Preiss and L. Zajíček, Stronger estimates of smallness of sets of Fréchet nondifferentiability of convex functions, Proc. 11th Winter School, Suppl. Rend. Circ. Mat. di Palermo, Ser. II, nr. 3 (1984), 219-223.

[Rain] J. Rainwater, Yet more on the differentiability of convex functions, Proc. Amer. Math. Soc. 103 (1988), 773-778.

[Ri₁] N. K. Ribarska, Internal characterization of fragmentable spaces, Mathematika 34 (1987), 243-257.

[Ri₂] —————————, A note on fragmentability of some topological spaces, C. R. Acad. Bulgare Sci. 43 (1990), 13-15.

[Ri₃] —————————, The dual of a Gâteaux smooth Banach space is weak star fragmentable, Proc. Amer. Math. Soc. 114 (1992), 1003-1008.

[R-V] A. W. Roberts and D. E. Varberg, Convex Functions, Academic Press, New York-San Francisco-London (1973).

[Ro₁] R. T. Rockafellar, Convex functions, monotone operators and variational inequalities, from Theory and Applications of Monotone Operators, Proc. NATO Adv. Study Inst., Venice, Italy, (1968), 35-65.

[Ro₂] —————————, Local boundedness of nonlinear monotone operators, Mich. Math. J. 16 (1969), 397-407.

[Ro₃] —————————, On the maximal monotonicity of subdifferential mappings, Pacific J. Math. 33 (1970), 209-216.

[Ro₄] —————————, Monotone operators associated with saddle-functions and minimax problems, in Nonlinear Functional Analysis, Part 1, F. E. Browder, ed., Proc. Symp. Pure Math., vol. 18, Amer. Math. Soc. (1970), 241-250.

[Ro₅] —————————, On the maximal monotonicity of subdifferential mappings, Pacific J. Math. 44 (1970), 209-216.

[S-P] J. Saint-Pierre, Sur le théorème de Rademacher, Sem. d'Analyse Convexe, Univ. des Sci. et Techn. du Languedoc, Montpelier (1982), Exp. Nr. 2.

[Sch₁] W. Schachermayer, For a Banach space isomorphic to its square the Radon-Nikodým property and the Krein-Milman property are equivalent, Studia Math. 81 (1985), 329-339.

[Sch₂] —————————, The Radon-Nikodým property and the Krein-Milman property are equivalent for strongly regular sets, Trans. Amer. Math. Soc. 303 (1987), 673-687.

[Si₁] S. Simons, The least slope of a convex function and the maximal monotonicity of its subdifferential, J. Optimization Theory and Applications 71 (1991), 127-136.

[Si₂] —————————, Les dérivées directionelles et la monotonicité des sous-différentiels, Sém. d'Initiation à l'Analyse (Sém. Choquet), Paris (to appear).

[Sm] V. L. Smulyan, Sur la dérivabilité de la norme dans l'espace de Banach, C. R. Acad. Sci. URSS (Doklady) N. S. 27 (1940), 643-648.

[Ste₁] C. Stegall, The duality between Asplund spaces and spaces with the Radon-Nikodym property, Israel J. Math. 29 (1978), 408-412.

114   References

[Ste₂]   _____, Optimization of functions on certain subsets of Banach spaces, Math. Ann. 236 (1978), 171-176.

[S-S]   K. Sundaresan and S. Swaminathan, Geometry and Nonlinear Analysis in Banach Spaces, Lect. Notes in Math., Nr. 1131, Springer-Verlag (1985).

[Tal₁]   M. Talagrand, Deux exemples de fonctions convexes, C. R. Acad. Sc. Paris 288 (1979), 461-464.

[Tal₂]   _____, Renormages de quelques C(K), Israel J. Math. 54 (1986), 327-334.

[Tay]   P. D. Taylor, Subgradients of a convex function obtained from a directional derivative, Pac. J. Math. 44 (1973), 739-747.

[Tr]   S. L. Troyanski, An example of a smooth space, the dual of which is not strictly convex, (Russian) Studia Math. 35 (1970), 305-309.

[Ver]   M. E. Verona, More on the differentiability of convex functions, Proc. Amer. Math. Soc. 103 (1988), 137-140.

[V-V₁]   A. Verona and M. E. Verona, Locally efficient monotone operators, Proc. Amer. Math. Soc. 109 (1990), 195-204.

[V-V₂]   _____, A note on minimal usco maps, Canad. Math. Bull. 34 (1991), 412-416.

[V-V₃]   _____, Characterizations of maximal monotone operators, Nonlin. Anal. - Theory, Methods & Applications (to appear).

[Ze]   E. Zeidler, Nonlinear Functional Analysis and its Applications, Vol II/A, Linear Monotone Operators; Vol II/B, Nonlinear Monotone Operators, Springer-Verlag (1985).

INDEX

INDEX OF SYMBOLS